Chaos in Switching Converters for Power Management

Enric Rodríguez Vilamitjana
Abdelali El Aroudi
Eduard Alarcón

Chaos in Switching Converters for Power Management

Designing for Prediction and Control

Springer

Enric Rodríguez Vilamitjana
Department of Electronics Engineering
Technical University of Catalunya (UPC)
Barcelona
Spain

Eduard Alarcón
Department of Electronics Engineering
Technical University of Catalunya (UPC)
Barcelona
Spain

Abdelali El Aroudi
Universitat Rovira i Virgili
Tarragona
Spain

ISBN 978-1-4939-0058-9 ISBN 978-1-4614-2128-3 (eBook)
DOI 10.1007/978-1-4614-2128-3
Springer New York Heidelberg Dordrecht London

Softcover reprint of the hardcover 1st edition 2013

Printed on acid-free paper

Springer is part of Springer Science+Business Media (www.springer.com)

Preface

Switching power converters are at the crossroads of a notable number of researchers from various disciplines, including circuit theorists, physicists, mathematicians, and power electronics engineers. This book is addressed to electrical engineering students and practicing engineers with interests and needs in the field of nonlinear dynamical behavior and bifurcation analysis and control of switch-mode power supplies. This book takes an engineering-oriented standpoint at bifurcation prediction in these circuits by combining different mathematical tools. The ultimate goal of this book is to explain how these phenomena can take place, that is, how to characterize, predict, and control their occurrence, in particular in the framework of progress lines in the field of power converters, namely for miniaturized and in particular integrated on-chip power management as well as for emerging applications such as wideband power amplifiers.

In contrast to other existing works in this field, nonlinear behavior is predicted by using simple design-oriented circuit-centric expressions which depend upon the circuit power stage parameters and which can explain consistently such behavior without necessarily resorting to the language of mathematical formulations widely used hitherto. The approach in this book is more like that of the systems engineer, hence considering the whole system as a set of well-defined subsystems, so that via an approximately top-down manner the effect of each subsystem upon the overall dynamical behavior is revealed.

The book begins in Chap. 1 with a short survey of switching power converters for modern power management architectures along with an introduction to dynamics and stability models of switching power converters. The review argues and instantiates that design-oriented models for predicting the complete (and rich) behavior of such circuits are missing, even for basic switching power converters.

Chapter 2 characterizes the dynamics exhibited by the converter circuits as a function of a complete parametric design space thereby revealing the different complex behaviors. The characterization is extended by evaluating the impact of the different dynamic modes upon the final application by means of defining a set of circuit-centric application-aware performance metrics.

In Chap. 3, the prediction of the boundaries between different dynamic behaviors is tackled from a design standpoint, starting from the simple hypothesis that the ripple magnitude at the PWM modulator can predict the boundary between the desired stability and nonlinear behavior. The hypothesis is validated with numerical simulation, experimental and discrete-time models results, and extended to different controllers. Maintaining the design-oriented standpoint, this book moves toward facilitating the synthesis and design of controllers oriented to avoid unstable dynamic modes by presenting a frequency domain perspective—a common working domain for power electronic engineers—of the conditions of occurrence of the various nonlinear behaviors. The model oriented to predict bifurcations is then used, complementary to classical linear modeling approaches, to carry out a complete system stability analysis in which both slow-scale and fast-scale instabilities are considered. The availability of such frequency domain interpretation is used in Chap. 4 for exploring strategies to design and implement new controllers which can alter the conditions of occurrence of period-doubling bifurcations. As a result from the development of such design-oriented tools, this chapter presents a set of controllers which are sensitive to implementation issues while they also consider other figures of merit such as the impact upon the complete stability boundaries or the ability to extend the miniaturization limits, set by area, efficiency, and ripple tradeoffs.

Finally, Chap. 5 explores the nonlinear dynamics of more complex switching power converter circuits, such as a multilevel converter, in which additional state variables are added, or a switching power converter operating in tracking conditions as a switching power amplifier in which the quasi-static time-invariant assumption is not valid.

Contents

Chapter 1
Introduction

Trends in portable applications such as mobile terminals for next generation communications proceed in the direction of increasing the computational load (voice and data communications) while concurrently reducing size and enhancing operating life time.

Both trends require to investigate on complex power management architectures and circuits (considered to be key enabling technologies), which make possible the development of new features according to the aforementioned trends to develop high-density and high-efficiency portable systems.

Figure 1.1 shows the system block diagram of a telecommunication battery-operated portable device, such as a mobile phone, where multiple subsystems can be identified, such as microprocessor, audio interface, display, baseband processing and Radio-Frequency (RF) subsystems. Such subsystems are properly energy supplied by means of different so-called Point of Load (PoL) voltage regulators, but there is also power management at system-level, which carries out energy management of the battery to achieve improved overall efficiency and therefore to improve the battery operation time.

The major power consumers that can be identified in the system are the Radio Frequency Power Amplifier (RFPA), baseband digital circuitry, display and analog circuitry.

It is foreseen that the next generation of portable devices will increase their energy consumption due to the increased functionality. This fact, along with the slower increase of battery capacity, yields to an increased need and interest on the power management subsystems, continuously pushed for more miniaturization and higher power density.

E. Rodríguez Vilamitjana et al., *Chaos in Switching Converters for Power Management*, DOI: 10.1007/978-1-4614-2128-3_1, © Springer Science+Business Media New York 2013

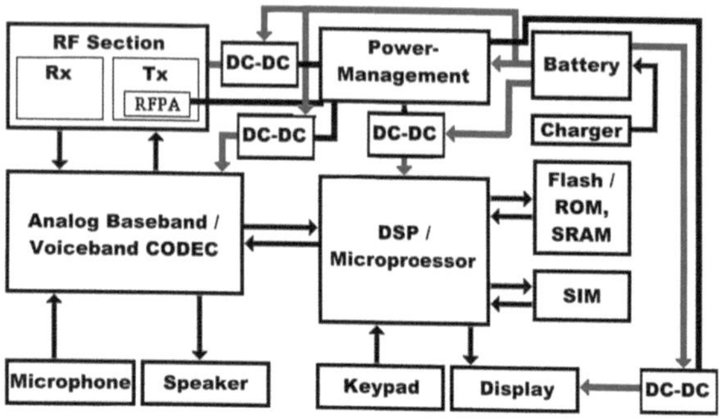

Fig. 1.1 Architecture of a battery-operated portable device (mobile phone)

1.1 Key Enabling Technologies for Power Management Subsystems

Trends in portable applications are associated with the improvement on integration and miniaturization of power electronics circuits along with their optimum power management control, hence being subsystems that guarantee the system efficiency in terms of energy and that have a high impact upon the system in terms of size. These trends are a line of convergence for the future systems in the field of portable and autonomous battery-supplied devices, where power management subsystems are one of the key factors that limits the system performance in terms of size and life time. It is possible to identify two trends in modern power management subsystems:

- Integration and miniaturization of power subsystems, which allow a system size reduction, increasing reliability and lowering cost.
- Advanced power management, taking into account system-circuit interactions (such as adaptive power supply of microprocessors or RF polar transmitters for digital communications).

Power subsystems, both in regulation and more advanced power management applications, are ideally based on switching power converters, paradigm of high efficiency power processing circuits. Such converters structure is composed of a set of reactive components (inductors and capacitors) along with switching devices (MOSFET, diodes), which by proper driving, control and modulation, allow to deliver efficiently the energy from the source to the load.

1.1.1 Integration and Miniaturization

The target scenario is a fully on-chip integration of the power management subsystem (circuit and control) together with the load circuits in the same silicon chip substrate resulting in a complete Powered System on Chip (PSOC).

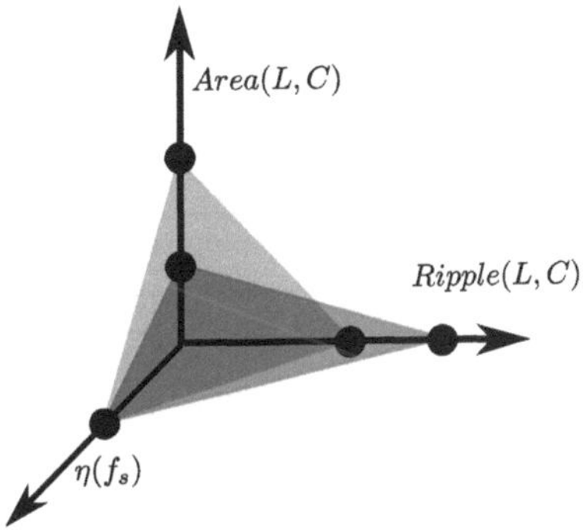

Fig. 1.2 Design-space of performance metrics limitations for integration: area, efficiency η and ripple. Two cases are illustrated: low area high-ripple and low-ripple high-area by keeping constant switching frequency

The common trend towards the miniaturization target is the size/value reduction of passive components. This is usually associated with integration since smaller passive components implies easier integration in MOS technology so that both miniaturization and integration techniques are usually set in the same group.

First steps made in integration techniques addressed the active parts Stratakos et al. (1994), Kursun et al. (2004) and Jung et al. (1999), such as MOSFET switch and drivers, while passive components were externally placed out of the integrated circuit. Subsequently, some works were carried out on passive components integration. As regards the inductor, the integration main drawback is due to the planar nature of MOS technology and the lack of ferromagnetic material in a standard process. With the evolution of MEMS processes some works have been carried out in the monolithic integration of the inductor, such as in Ahn et al. (1996), Iyengar et al. (1999). Other techniques are based on planar CMOS inductors and others on bond-wire based inductors Mohan et al. (1999). As for output capacitance integration, the most common alternative relies on the parallel plates approach Hajimiri (2002).

The circuit-level design-space aimed to integration is limited by three performance metrics Villar et al. (2003): area, efficiency η and ripple. Then, reducing the inductance and capacitance parameter values result in a decrease of the area, but also increases the natural frequency of the implicit low-pass filter of the DC/DC converter, so that the ratio between the switching frequency and the natural frequency decreases, hence yielding to high ripple. In order to increase such ratio and reduce the ripple, which is limited by the load specification, the switching frequency is usually increased, but this implies increasing the switching losses, hence decreasing the overall converter efficiency. The tradeoff between these three performance metrics is shown in Fig. 1.2

It is worth mentioning that the converter performance can also be measured from other complementary performance metrics, such as transient response settling time due to a load change, but this will not only depend upon converter parameters but also upon the controller and modulator strategy.

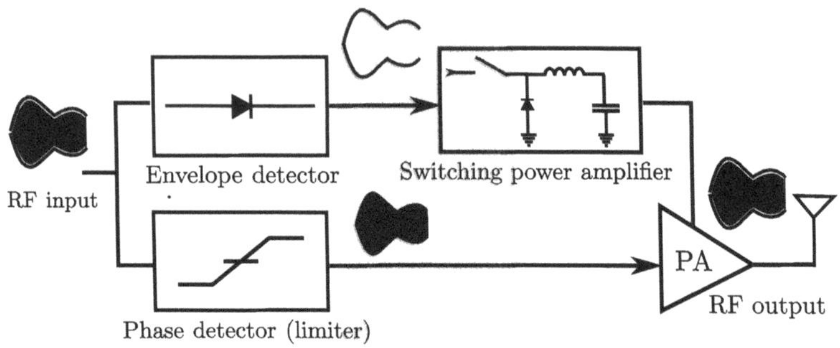

Fig. 1.3 EER technique including wide band switching power converter as adaptive power supply

1.1.2 Advanced Functionality Power Management Circuits

Switching power converters have hitherto mainly been used in regulation applications, where it is required to supply a constant voltage to a load. This regulation function requires a fast reaction in front of a load step change, which demands to investigate new topologies and control techniques to improve transient response, and therefore bandwidth, to meet the stringent requirements of the new generation power loads (such as GHz-clocked microprocessors).

A further complexity has been added in the functionality of the switching power converters, using them as trackers of wide band time-varying reference signals. This is the case of RFPA, in which the necessity of increasing communication rates implies to use more complex modulation strategies established by communications standards. These modulation strategies use the spectrum more efficiently by using non-constant envelopes, requiring a high-density modulation constellation, such as OFDM, QPSK or CDMA, among others. The necessity of using non-constant envelope does not allow RF power amplification by a high efficiency RF switching amplifier (class E/F), because of its nonlinear amplitude response, but by means of linear amplifiers (class A) which achieve very limited maximum efficiency.

The interest in developing new wide band adaptive power management circuit comes from the necessity of implementing high-efficiency high-linearity RF transmitters.

One architecture proposed long ago to wide-band power supply RF amplifiers is the so-called Envelope Elimination and Restoration (EER) technique Raab and Rupp (1994). It separates digitally-modulated complex signals into envelope and phase, and after efficient power amplification of both polar paths, they are combined before transmitting the signal to the antenna. The envelope is tracked and amplified by a switching power converter, based on buck-based switching converter, as it is shown in Fig. 1.3.

Although switching power converters used as wide band envelope trackers inherently result in tracking errors (both in-band and out-of-band-errors), higher switching frequencies leads to reduced tracking errors. Nevertheless, high switching frequen-

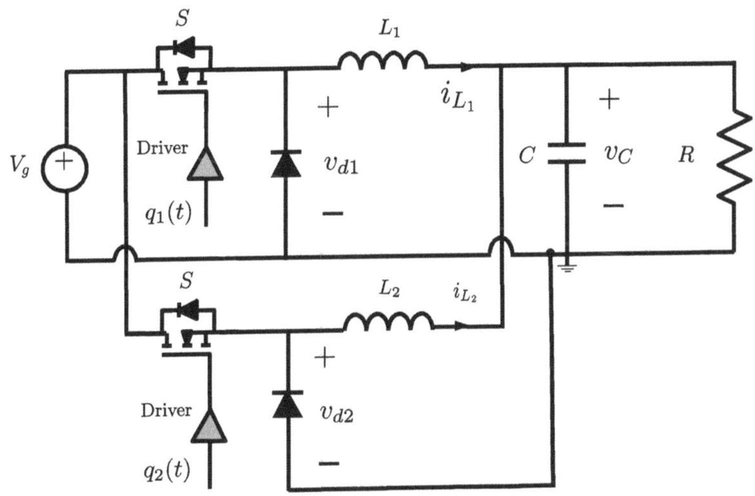

Fig. 1.4 Parallel connected (interleaving) topology based on buck converter

cies also yield to an increase of switching losses, resulting in an error-efficiency tradeoff.

1.2 Modern Power Management Approaches

In order to make feasible both aforementioned trends, key enabling technologies have been developed, composed of more complex architectures approaches that take into account different topology, control and functionality considerations.

1.2.1 Non-Conventional Switching Power Converter Topologies

This section attempts to review the main non-conventional topologies, usually derived from classical conventional ones Erickson and Maksimovič (2001), used with the aim of achieving better ripple, faster-response and/or smaller passive components. Most of the reviewed topologies in this section, and further in this book work, are based on a step-down buck converter, since within the portable battery-operated low-power scenario the supply voltage, such as a 4.2 V Lithium-Ion battery, is usually higher than the required load voltage, for instance a digital CMOS load microprocessor (around 1 V).

One of the most common topology is the interleaving converter Stanley and Ronald (1995), which is widely used in high-current low-voltage voltage-regulator module (VRM) applications among others. The topology is based on using multiple parallel-

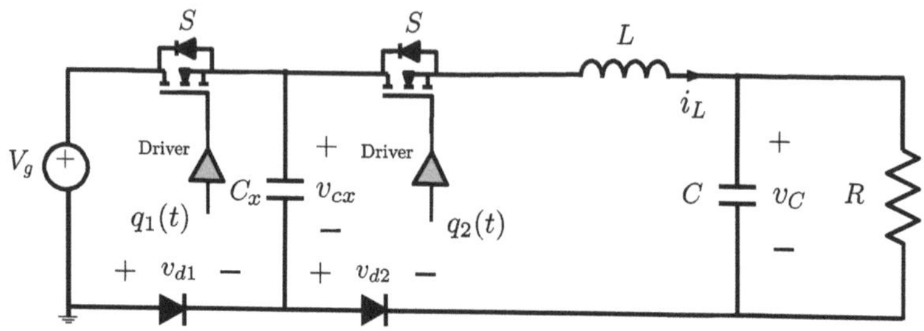

Fig. 1.5 Multilevel (2-cell, 3-levels) topology based on buck converter

connected converter topologies as it is shown in Fig. 1.4 for two connected converters. Its main advantages are its wider bandwidth tracking and lower output ripple compared to single converter topology whereas, on the other hand, it has complex control and higher implementation area.

Another less common non-conventional topology is the multi-level converter, shown in Fig. 1.5. They have been historically used for high-voltage DC–AC applications, due to both reduced blocking voltages and improved distortion performance Meynard et al. (2002). By properly controlling the switches, the topology could achieve constant (regulation) or non-constant (amplification) output waveform with low output ripple.

Other more advanced topologies have been developed as hybrid solutions, as it is the case of the Linear-Assisted (LA) topology Yousefzadeh et al. (2006a) in which the switching converter is used usually for high efficiency, despite it exhibits a nonlinear switched response, besides the ripple. On the other hand, linear regulators are very accurate and wideband but not efficient. In the LA topology, the is used to process the major part of the energy, whereas the linear regulator is used to reject the ripple and upper-band of the signal spectrum, thereby is only managing a small part of energy, avoiding a severe impact on the overall efficiency.

1.2.2 Non-Conventional Control

The control architecture consists of generating the different driving signal (depending upon the feedback state variable), the compensator, and the modulation strategies.

Conventional control architectures are based on either voltage-mode control (VMC) or current-mode control (CMC) along with using a high DC-gain amplifier combined with low pulse width modulation (PWM) gain, as it shown in Fig. 1.6. It uses fixed switching frequency with variable duty cycle and the control signal is obtained by comparing a saw-tooth waveform with the control signal proceeding from the error amplifier.

PWM VMC controller advantages are that they can be used in most converter topologies and exhibit a good DC regulation. On the other hand, they have slow

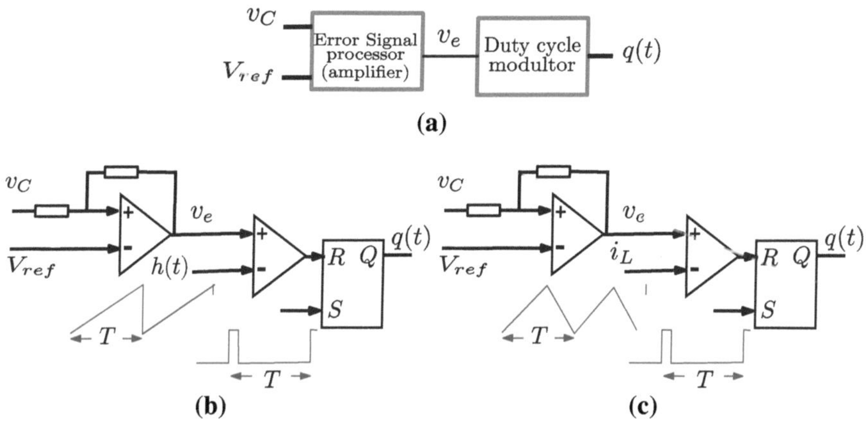

Fig. 1.6 **a** Conventional PWM modulator scheme **b** VMC and **c** CMC

response to input voltage or load perturbations, and require precise feedback dynamics compensation.

Another common technique to control the converter is CMC, in which the inductor current is used as feedback variable and acts as saw-tooth signal modulator Deisch (1978). With this technique, faster response is achieved by using inductor current instantaneous waveform. This technique is usually complemented with an outer voltage-feedback loop, as it is shown in Fig. 1.6, in order to regulate the output voltage according to load demands.

In a similar way as in CMC, in which the inductor current is directly feedback to the comparator, there is a variety of techniques, so-called ripple regulators Wester (1990) and Redl and Sun (2009), based on not using error amplifiers on the feedback path but high-gain duty-ratio modulators. These instantaneous control techniques do not require feedback dynamics compensation and have faster response to load and input perturbations by using the switching information given by the output ripple. On the other hand, depending upon its implementation, the switching frequency can not be easily determined, which makes difficult the components design, among other drawbacks such as potential exhibition of instabilities exhibition or poor DC-regulation.

All aforementioned control methods are based on analog domain (time-frequency) design, but there also exists the possibility of implementing these controllers in the discrete-time domain, yielding to more reconfigurable systems, but with added nonlinear dynamics because of the analog-to-digital conversion.

The feasibility of high-frequency, high-performance digital controller DC-DC applications started to be considered in the last decade. Based on custom architectures and microelectronic realizations of the key building blocks (see Fig. 1.7), including high-resolution high-frequency digital pulse-width modulator (DPWM)), simplified discrete-time compensators scheme ($G_c(z)$), and analog-to-digital A/D converters, such controllers offer the advantage of lower sensibility to parameter

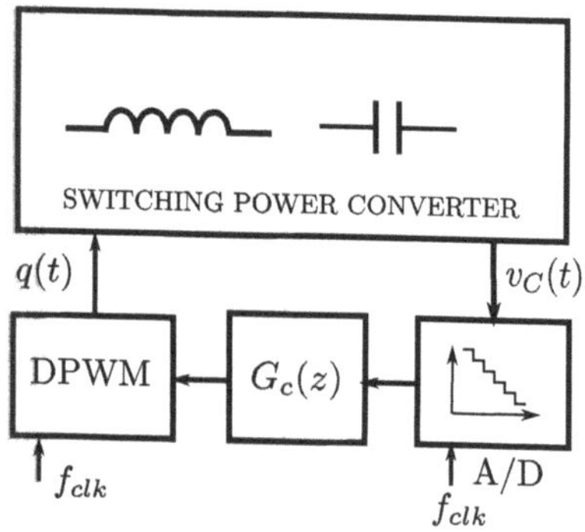

Fig. 1.7 Block diagram of a digital controller

variation, programmability, and reduction and elimination of external passive components, without compromising dynamic performance, simplicity and cost.

First works on digital control in a boost converter via a digital-signal processor (DSP) was presented in Tse and Tam (1994), obtaining however a low dynamic performance converter. Its switching frequency was around 3 kHz, restricted by the DSP operating frequency, number of instructions and delay. While that paper attempted to explore digital control with not so good regulation, in Dancy and Chandrakasan (1997) an ultra low power controller is reported, achieving a 88 % efficiency and switching frequency of 330 kHz for a buck converter. Subsequently in Prodic et al. (2001) a complete design implementation is presented with more complex control functions at 1 MHz switching frequency. A different DPWM approach has been studied in Syed et al. (2004) in order to improve the performance (delay, resolution and silicon area), while in Corradini and Mattavelli (2008) the effect of multi-sampling technique in the A/D converter is analyzed, leading to increasing the bandwidth of the converter.

The digital framework adds further complexity in the dynamic performance of converters by discretization of the feedback variable (through A/D converter) and the control PWM signal (through DPWM) along with adding additional nonlinearities to the converter, i.e. the sample-and-hold effect of the A/D converter. A notable effort has been done to establish stability criteria in Peterchev and Sanders (2003) and Peng et al. (2004) considering DPWM and A/D converter resolution. Furthermore, in Maksimovič and Zane (2007) a discrete-time model is derived exploring the effect of discretization upon the frequency response.

1.2.3 Functionality: From Regulation to Tracking

Power switching converters have usually been used as regulator circuits which have a fixed output voltage/current. Non-regulation applications have the aim of dynamically supplying energy to the load following a time-varying reference (tracking).

A conventional application where non-regulation switching power converters are considered, is in AC–DC or DC–AC converters, due to the necessity of shaping the input current in Power Factor Correction (PFC) circuits or output voltage (inverters) respectively, at very low mains frequency (50 Hz). Hence switching power converters act as a quasi-static signal tracker.

In the early 90's, the interest in switching converters grew in audio amplifier applications. Several companies and laboratories started to investigate on such topic, taking advantage of the previous knowledge about switching converters on regulation applications and their properties. As for audio amplifiers, the switching power converters have to amplify the audiofrequency range (from DC to 20 kHz).

So far, in the switching amplification field, all studies have been restricted to low frequency (audio bandwidth). As it was mentioned for the key enabling technology of RF transmitters with fast adaptive power supplies, in Sect. 1.1.2, interest is nowadays centred on wider bandwidth frequency range. This implies following a high bandwidth signal in order to amplify it, namely in the order of MHz (3G) or tens of MHz (WLAN).

Bandwidth analysis to produce a determined error on output signal by using EER topology based on the buck converter is suggested in Marco et al. (2006).

1.2.4 Modern Power Management Architectures Case Examples

Different applications considered combination of non-conventional functionality, topology and control revisited in the previous section. This section carries out a review of recent proposed architectures, based on switching converters, targeting performance improvement toward miniaturization and integration as well as power amplifier functionality.

A first example of a non-conventional approach in the RF adaptive power supply power amplifier field is given Yousefzadeh et al. (2005); Rodriguez et al. (2009); Vasic et al. (2009) in which a three-level buck-based converter is proposed as an RF power amplifier power supply. In both cases, the control is implemented digitally. Furthermore, the combination of RF power amplifier along with digital control for enhanced dynamics is studied in Yousefzadeh et al. (2006b) where the digitally controlled buck converter is used as RFPA supply.

Complementarily to efforts to reduce passive components in DC-DC converters by using different material/structure technologies (revisited in Sect. 1.1.1), there are some approaches at converter topology level that facilitate integration while preserving other performance metrics. For instance, in Schrom et al. (2004) a multi-phase converter is controlled by a hysteretic (ripple) controller in order to reduce passive components while improving transient response. In a similar way, in Corradini et al. (2009), a digital hysteretic controller is used in order to improve the system resolution.

Furthermore, in Villar and Alarcon (2008) a multilevel converter is used for full monolithic integration of a DC-DC converter taking advantage of its low ripple performance, that allows to reduce the inductor size. The flexibility of digital control has been used in Bergveld et al. (2008) to implement a simple zero-voltage switching (ZVS) technique, in order to improve efficiency and to facilitate the converter integration.

The modern power management architectures presented in this section imply the use of advanced topologies, control and functionalities, hence increasing the complexity of their dynamical analysis. The next section revisits the different dynamic behaviors that can be exhibited by a switching power converter and discusses the available stability models to predict the boundary between such dynamic behaviors.

1.3 Dynamics and Stability Models of Switching Power Converters

The previous review was focused on the state-of-art of key technologies pursuing to allow future power subsystems requirements. With this aim in sight, it is important to characterize in terms of dynamics these power management circuits in order to guarantee their proper dynamic behavior and performance.

Switching power converters, due to their switching nonlinear nature, are prone to exhibit diverse types of nonlinear phenomena.

1.3.1 Overview of Nonlinear Dynamical Systems

Roughly speaking, any system that has a set of independent state variables and a deterministic relation between them and the system's inputs, constitutes a dynamical system. Such dynamic system can be described mathematically as:

$$\dot{\mathbf{x}}(t) = \mathbf{f}(\mathbf{x}(t), \mu) \tag{1.1}$$

where $\mathbf{x}(t)$ is the evolution in time of the state-variables, $\mathbf{f}$ is the vector field, and μ is a vector of external system parameters. If the vector field $\mathbf{f}$ does not depends upon time, the system is called *non-autonomous*. Otherwise, the system is called *autonomous*. Examples of non-autonomous systems are classical fixed-frequency (PWM) power converters while free-running converter such as hysteretic converters are autonomous systems.

In such systems, so-called deterministic systems, the exact way that the state variable follows can be determined by the vector field. For linear systems, given an initial condition to state-variables, the solution, also known as *trajectory*, is completely determined and a closed-form solution can be found.

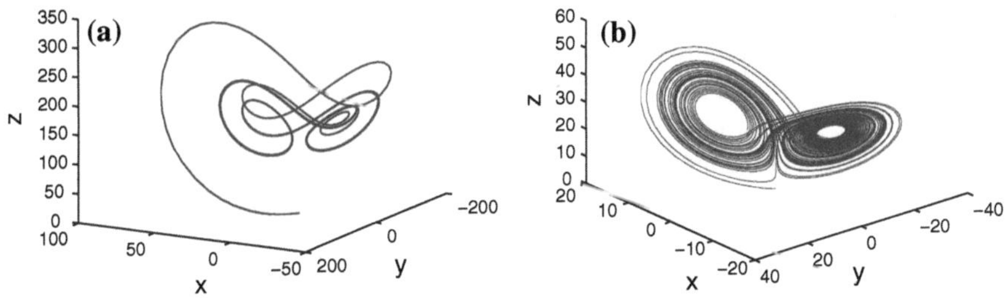

Fig. 1.8 Trajectories from the Lorenz system $\dot{x} = 10(y-x)$, $\dot{y} = x(r-z)$ and $\dot{z} = xy-3z$. **a** Limit cycle in steady-state with r=160 **b** Chaotic attractor in steady-state with r=25. Initial conditions x=0.1, y=0.1 and z=0.1

However, in nonlinear systems, the development of trajectories and the closed-form solution requires of higher complexity analysis. The system, after an initial transient, enter its steady-state. This steady-state solution to which the system converges is called an *attractor*. In nonlinear systems, the convergence to an attractor will depend upon the initial condition of state-variables. Thus, to determine the steady-state behavior of the system, it is necessary to know the possible attractors:

- *Equilibrium point*: The steady-state solution is a point in the state-space.
- *Limit cycle or a periodic orbit*: The steady-state trajectory moves along a closed path in the state-space.
- *Quasi-periodic attractor*: The steady-state trajectory moves on the surface of a torus.
- *Chaotic attractor*: Although there is no universally accepted mathematical definition of chaos, a commonly used definition says that, for an attractor to be classified as chaotic, it must have the following properties:
  - the corresponding system must be sensitive to initial conditions
  - it must be topologically mixing which means that the system will evolve over time so that any given region or open set of its phase space will eventually overlap with any other given region; and
  - its periodic orbits must be dense which means that every point in the space is approached arbitrarily closely by periodic orbits.

Examples of a periodic orbit and a chaotic attractor are shown in Fig. 1.8, obtained from numerical simulation of the Lorenz system.

If parameters in a nonlinear system are varied, then the system may abandon one solution and become attracted to another one. This phenomenon, namely the sudden change of qualitative behavior when a parameter is changed, is called a bifurcation. Bifurcations can be classified according to the type of qualitative change that takes place.

- Period-doubling or flip bifurcation: This type of bifurcation is characterized by a sudden doubling of the period of the stable periodic orbit.

- Hopf bifurcation: This type of bifurcation is characterized by a sudden birth of a limit cycle from an equilibrium point. In discrete-time systems, this is also called a secondary Hopf or a Neimark-Sacker bifurcation.
- Saddle-node bifurcation: This type of bifurcation is characterized by a sudden appearance/disappearance of a pair of equilibrium points. In discrete-time systems, other names for this phenomenon are tangent or cyclic fold bifurcation.

The aforementioned types of bifurcation do not involve structural changes and they are called *smooth bifurcations*. However, there are other instabilities that occur due to a structural change in the system, called *border collision bifurcation* or non-smooth bifurcations.

Considering the described possible bifurcation types and equilibrium points, different behaviors can be exhibited by varying a parameter. Depending on the precursor state, different routes to chaos can be distinguished:

- *Route to chaos via period-doubling*: as one parameter is modified, the system may undergo a period-doubling bifurcation, which is repeated as the bifurcation parameter is moved further and eventually ending up in chaotic behavior.
- *Route to chaos via quasi-periodicity*: as a parameter is modified, the system may undergo a Hopf bifurcation and by continuing varying such parameter the periodicity of the limit cycle is also modified.

1.3.2 Switching Power Converters Dynamics and Modeling

Switching power converters can exhibit the different kinds of instabilities introduced in the previous section due to their nonlinear behavior. The most well-known, Hopf bifurcation or also called slow-scale instability () SSI within the switching power converter context, leads to exhibit low frequency oscillations El Aroudi et al. (1999). However, switching power converters can also exhibit other kinds of instabilities. Exhibition of chaotic instabilities and subharmonic oscillations in switching power converters, called fast-scale instabilities (FSI), have been reported in the early 80's in Redl and Novak (1981) for the case of a buck converter under CMC. The SSI and FSI terms were first introduced in the power electronics field in Mazumder et al. (2001).

The converter state variables, namely inductor current and capacitor voltage, and the control signal at the modulator input as a function of time, along with the spectral domain and phase portrait of each attractor are shown in Fig. 1.9.

The first observations of FSI in switching power converters were just justified as instabilities. It is later where the concept of chaotic behavior or subharmonic! oscillation came from the Physics discipline, such as in Deane and Hamill (1989) for the case of buck and boost topologies. These first works, based on mere observations and experimental corroboration Wood (1989); Krein and Bass (1990), opened a new branch of dynamic models of switching power converters able to predict all the dynamics of the converter.

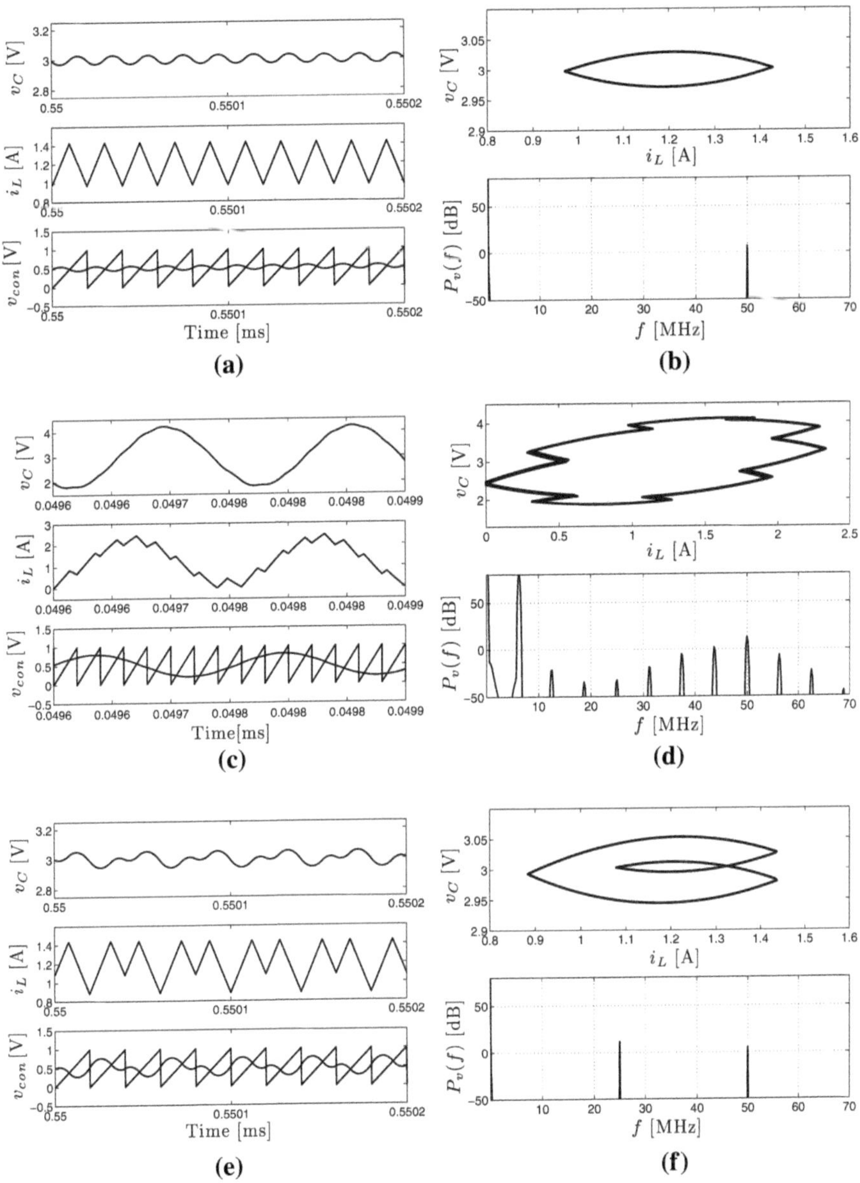

Fig. 1.9 Capacitor voltage, inductor current and modulator input waveforms along with phase portrait and frequency domain representation, being $P_v(f) = 20\log(|V_C(f)|)$, in a VMC buck converter with PI control for the occurrence of different attractors (**a**, **b**) period-one (**c**, **d**) low frequency oscillation (SSI) and (**e**, **f**) period-two bifurcation (FSI)

The modeling and prediction of the aforementioned instabilities in switching power converters, necessary to guarantee the stability of the design (or define robustness of the system), have been tackled from different standpoints during the last decades. On the one hand, obtaining an accurate model of the system, which implies the use of more sophisticated mathematical tools, hence losing simplicity and the circuit standpoint. On the other hand, looking from a design and circuit perspective, allows simplicity as well as proper synthesis of the controller, but assuming certain inaccuracy. Both trends were set out by Middlebrook and his coworkers Packard and Wester in the late 70s. The first one develops discrete-time models Packard (1976), which require to abandon the continuous-time domain, and somehow lack a design-oriented perspective but can accurately characterize the switching power converter dynamics. The second model is based on a time-averaged power stage model Wester and Middlebrook (1972), which is time-invariant and design-oriented, but fails to predict the occurrence of FSI. Both works are considered the origin of two trends in switching power converter dynamic models, although the first one has been later biased to approximated models, so-called sampled-data models, that allow a frequency domain understanding which is more related to a design-oriented perspective. This section presents a review of both the averaged and discrete models. The state of art shows models from basic switching converters to more advanced topologies as presented in the previous section.

1.3.3 The Discrete-Time System

The discrete-time model of switching power converters was first presented in Packard (1976). The model is based on concatenating the evolution of state variables during each switching subinterval, as it is shown in Fig. 1.10, in order to accurately obtain the evolution of the state variables during the switching period.

$$x_{n+1} = \mathbf{f}(x_n, \mu) \tag{1.2}$$

where x_n is the state variable vector, $\mathbf{f}$ is the mapping function and μ is the external parameter vector.

From the discrete-time model, the stability analysis is carried out by studying the local (linearized) behavior in the vicinity of the steady-state thereby extracting the Jacobian matrix, whose eigenvalues (λ) allow to identify the type of instability occurrence depending on their value when they cross the unit circle: if they have imaginary term, the Hopf bifurcation, so-called Neimark-Sacker bifurcation in discrete-time systems, occurs. If a crossing of one eigenvalue occurs at -1 period-doubling bifurcation, so-called flip bifurcation in discrete-time systems, occurs, and if crossing of one eigenvalue occurs at 1 saddle-node bifurcation takes place. These three categories are shown in Fig. 1.11.

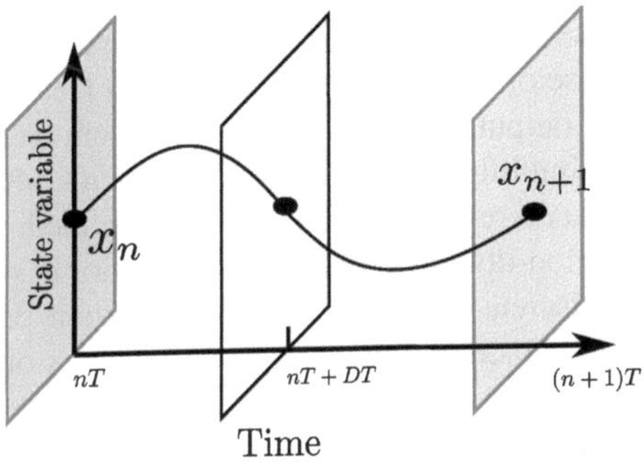

Fig. 1.10 Discrete time model illustration by sampling a two configuration system with period T

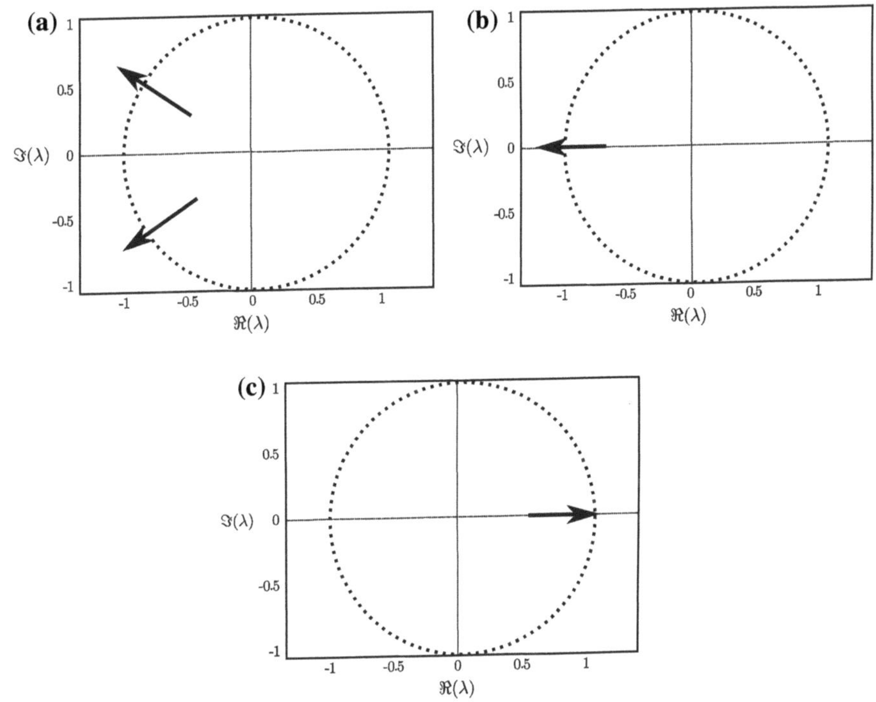

Fig. 1.11 Classification of the different smooth bifurcations of discrete-time systems by the way the eigenvalues (λ) cross the unit circle: **a** Neimark-Sacker bifurcation occurrence **b** period-doubling bifurcation and **c** Saddle-node bifurcation

While the first authors working in this area, using the sampled data model, which will be reviewed in Sect. 1.3.3.2 came from an engineering background, almost one decade later, similar models were re-developed from authors coming form the non-linear dynamics community, giving a mathematical standpoint, without referring to previous aforementioned works. This perspective was first reported by Hamill and Jeffries (1988) for the case of a CMC buck converter, in Deane and Hamill (1990) for a first order (without output capacitor) VMC buck converter and in Deane and Hamill (1991), where a simplified model is proposed for a CMC buck converter. In Hamill et al. (1992) a general review of the model is presented. These works are the first ones to talk about period-doubling and chaotic behavior in switching converters.

Works dealing with discrete-time models continued their development because of the necessity to fully characterize the complete dynamics of more sophisticated switching power converters.

The first discrete model, based on a linear ripple assumption, is presented in Deane (1992) for the case of a Boost converter with CMC. Subsequently, it is in Tse (1994a) where a buck converter is addressed in DCM. It is relevant to note that these first studies are made in CMC or DCM, since in these two case, the system is one-dimensional.

The first analysis of a VMC buck working in CCM and under a proportional gain is carried out in Fossas and Olivar (1996). Later, the VMC buck converter with a PI compensator is analyzed in Papafotiou and Mar-garis (2002). While a VMC buck with proportional gain only exhibits FSI, with a PI compensator, it could also exhibit SSI, yielding to possible interaction between both kind of instabilities. This interaction is also analyzed in Mazumder et al. (2001) for a buck converter with a second-order input filter.

As regards the boost converter under VMC, in Tse (1994b) it has been analyzed working in DCM by means of a one-dimensional discrete-time model. Later in El Aroudi et al. (1999), Toribio et al. (2000) the stability regions are obtained, proving, by numerical simulations and measurements, that the converter only exhibits FSI instability in DCM.

Further improvements of the models were developed by adding more realistic parameters taking into account parasitic effects. It is the case of Banerjee and Chakrabarty (1998), where a model including parasitic components is analyzed. The authors also present a design parameter space exploration encompassing input voltage and load variation, which are of special interest for regulation applications. In Parui and Banerjee (2003), the stability has been investigated for the converter operating mode changing between CCM and DCM as a result of load fluctuations. Subsequently, in Banerjee et al. (2004) the inherent switch delay of a power converter is taken into account in the discrete-time model.

A summary of analysis for nonlinear phenomena by means of a discrete-time model for buck, boost and buck-boost topologies has been reported in di Bernardo et al. (1997) and a new asynchronous discrete-time model is developed in di Bernardo et al. (1998). From the analysis of such map, the authors find a closed-form expression for period-doubling condition and point out that this bifurcation is related to the value of the derivative of the control signal at the switching instant. Subsequently, further elaboration of discrete-time models for predicting bifurcations and chaos in DC-DC switching power converters are presented in di Bernardo and Vasca (2000), Banerjee et al. (2000a), Banerjee et al. (2000b) and El Aroudi et al. (2005).

Most of the works related to the stability analysis in switching power converters have hitherto been focused in a basic regulation application with conventional topologies. The next paragraphs discuss more complex architectures as discussed in Sect. 1.2.

1.3.3.1 Stability of Advanced Power Management Architectures and Applications from Discrete-Time Modeling

In Tse et al. (2003a) a first work was presented where a slow variation of the input voltage was considered. The work presents a PFC stability analysis based on parametric sweep by varying the input voltage amplitude and phase.

Subsequently, a varying reference is added by coupling a spurious signal in Tse et al. (2003b) yielding the converter to exhibit of a phenomenon so-called intermittent chaos. Moving borders of the PWM modulator have been considered in Ma et al. (2004). Recently in Dai et al. (2007), El Aroudi et al. (2008a) SSI is studied in single-stage isolated PFC converters.

As regards non-conventional topologies discussed in Sect. 1.2.1, there are various works for the case of interleaving-converters. Chaos exhibition in these systems is reported in Iu and Tse (2000) for the case of parallel connected master-slave buck converters and in Batchvarov et al. (2002) for the case of interleaving boost converters. Furthermore, in Iu and Tse (2005) an interleaved buck under CMC is explored, proving that the period-doubling phenomenon is not exhibited in this case. Further analysis upon boundaries between SSI and FSI have been carried out in Huang et al. (2007) for an interleaved buck-based converter by means of a discrete-time model.

As regards to multilevel or multicell topologies, they have only been recently analyzed by developing discrete-time models in El Aroudi and Robert (2006) and El Aroudi et al. (2008b).

In non-conventional control methods, there exist few works related to discrete-time models. Most works have been focused in CMC Verghese et al. (1989). Recently in Redl and Sun (2009) different types of ripple controllers have been reviewed and their stability has been analyzed. Besides, in Cal- vente et al. (1996) sliding-mode controlled boost converter is analyzed under PI compensator. Subsequently, in Magauer and Banerjee (2000) hysteretic control is applied to an AC inverter and its stability is analyzed by means of discrete-time models and experimentally validated.

Discrete-time models have been recently naturally reconsidered in digital control, due to their inherent discrete-time nature via the use of z-transform, very common in the domain of digital design. In Van de Sype et al. (2004) a frequency domain model is given in the z-domain for a first-order buck converter, starting from modeling the effect of different modulators. Besides, in Maksimovič and Zane (2007) a new small signal model of digitally controlled DC-DC converters is given, taking into account sampling, modulator effects and delays in the control loop. The model is based on obtaining discrete-time recurrence equation and then transforming it into the z-domain. In addition, simpler models have been previously found in Maity et al. (2007), which derives an exact discrete model for the case of a buck with a PWM-1 modulation which is based on adding a sample and hold module in the feedback loop before the PWM modulator. The work demonstrates the different types of border collision bifurcation that may occur in the system.

1.3.3.2 From Discrete-Time Models to Frequency Domain Representation: Sampled-Data Models

While the discrete-time models accurately capture the dynamics of the converter, their main drawback is the lack of a design-oriented standpoint that facilitates the proper design of controllers to avoid undesired unstable behavior. With this aim in sight there have been some attempts to transform and simplify such discrete-time models into the frequency domain.

As it was mentioned before, the discrete-time model was extended by mixing both the averaged and the discrete models, yielding to the so-called sampled-data models Brown and Middlebrook (1981), that combines the continuous form of the averaged state equations and the accuracy of the discrete-time model. It is based on adding the sampling effect due to the pulse-width-modulator in the context of averaged small signal models. Subsequently, there appeared more elaborated studies on sampled-data models in Verghese et al. (1986, 1989). Following this approach, a new continuous-time frequency model was obtained for a buck converter under CMC in Ridley (1989) through approximating the discrete-time model. The work obtained an s-domain transfer function, shown in Fig. 1.12, that approximates the sampling action thereby allowing to obtain the control-to-output transfer function, including the effect of the compensating ramp, which is widely used in CMC controllers to control instabilities. It is worth to note that the work concludes that the instability may occur, even for duty cycles D lower than 0.5 ($D<0.5$) due to the outer voltage-loop (in general, it is assumed that under pure CMC, FSI only is exhibited for $D>0.5$).

A further generalized approach for modeling CMC converters has been proposed in Tan et al. (1995) and a similar development has been carried out in Bryant and Kazimierczuk (2005) for the case of boost under peak-CMC. In all cases the sampling effect of the discrete-time model is simplified in the s-domain taking advantage of the sample and hold nature of CMC.

Limits of such models have been explored in Pavljasevic and Maksimovič (1997) while in Sun et al. (2000a) the link between discrete-time and continuous time models is explored for different transformations from the z-domain to the s-domain.

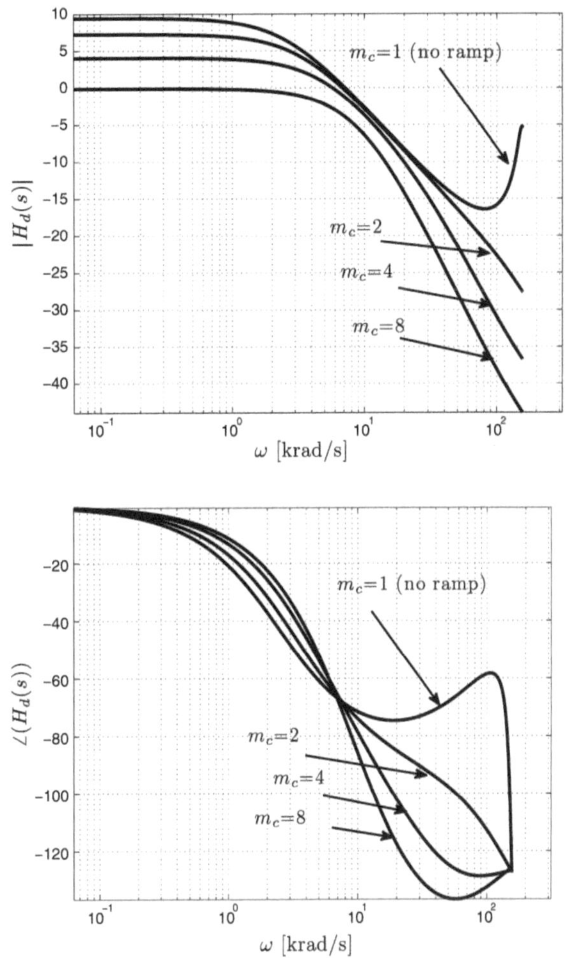

Fig. 1.12 The control-to-output transfer function $H_d(s)$, in the frequency domain, considering the effect of the compensating ramp m_c obtained from the expression and values proposed in Ridley (1989), $D = 0.4545$, $L = 37.5\,\mu$H, $C = 400\,\mu$F, $R = 1\,\Omega$ and $f_s = 50$ kHz

1.3.4 Design-Oriented Circuit-Based Models: The Average Model

The models presented in the previous section characterize accurately the converter dynamics and can predict all instabilities that can be exhibited. However, these models lack a design-oriented use that facilitates the stability analysis and quantifies easily propensity the converter is to exhibit instabilities.

The most common design-oriented model is the averaged model proposed in Wester and Middlebrook (1972). It is based on averaging the converter waveforms directly, whereby all manipulations are performed on the circuit diagram and hence it gives a more physical interpretation. The model is based on replacing the converter switches (MOSFET and diodes) by voltage and current sources, to obtain a time-

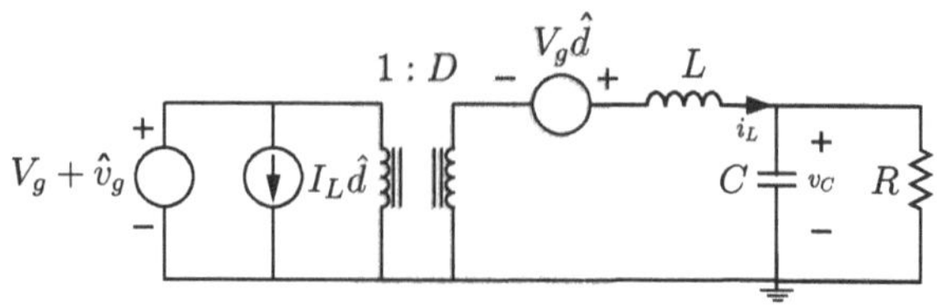

Fig. 1.13 Small-signal averaged model of buck converter

invariant circuit topology, and then obtain a small-signal linearized model, as it is shown in Fig. 1.13 for a buck converter.

Subsequently, a model based on state-space averaging approach was developed in Middlebrook and Cuk (1976), giving a unified approach to modeling DC-DC converters. Both models achieve the same small-signal equivalent circuit by two equivalent ways, although the first one has a more circuital standpoint. These models allow the application of Laplace s-domain analysis yielding to obtain simple stability conditions by means of Bode frequency domain representations or Nyquist plots.

The main problem of the average model is that it is based on an approximation, due to its intrinsic averaging process, and then it can only predict low frequency oscillation lower than one half of the switching frequency so that it does not include switching frequency information.

Posterior improved averaged models try to solve the shortcomings of the original model, especially focusing on switching information which allows to improve the model accuracy and eventually predict FSI. Note that the simplification of such sampling information and translation into the s-domain was the main objective of the work shown in previous Sect. 1.3.3.2.

In Lehman and Bass (1996) the switching frequency effects were considered by adding periodic ripple functions to the classical averaged model for the case of a boost converter under CMC. By adding such periodic ripple functions, the model provides more accurate transient and steady-state responses.

Most of the works presented before are focused in improving the model for the CMC case (this trend is also observed in discrete-time models) since it is only composed of one state variable hence allowing a simpler analysis. However, there is a lack of models that address the prediction from a design-oriented standpoint for VMC converters.

So far, design-oriented tools conceived to predict subharmonic oscillation are based on a mere design-space exploration such as in Chakrabarty et al. (1996) or Cheng et al. (2003). In a similar way, in Wu et al. (2005) this exploration has been carried out for a single stage PFC converter.

A complementary approach for predicting FSI occurrence is presented in Fang and Abed (2001), which consists on the derivation of Fourier series in both period one and period-two regimes. From such analysis, it was possible to derive a Fourier series-based closed-form inequality for predicting the stability boundary, but the analysis requires certain complexity and lacks a design-oriented standpoint.

This section has revisited the different approaches for modeling the dynamics of switching power converters. Whereas most of the approaches have been done though discrete-time formulation, thereby obtaining an accurate model of the system but missing a design-oriented standpoint, the averaged model, which can be derived from a circuit-based approach and gives a design-oriented standpoint, has an important limitation in predicting FSI. The intermediate solution from which it is possible to obtain frequency domain models, namely sampled-data models, still lacks a circuit-based approach and is mainly limited to CMC.

The section unveils a gap regarding the prediction of FSI not only in advanced architectures but also in classical VMC. It is worth to note that the VMC is widely used in regulation application, even in CMC, due to the necessary output regulation.

1.4 Chaos Control Methods

The previous section has carried out a review of the state of the art of different models to predict instability of different topologies aiming power management applications. However, as it has been observed, most of models, especially in VMC and other more advanced architectures are addressed from a discrete-time perspective, hence lacking a design-oriented standpoint and therefore challenging the synthesis of proper controllers.

The research on control methods not only to reject chaotic behavior, but also to induce it, so-called anti-control Morel et al. (2004), has been a focus of study in recent years.

The interest in anti-control, comes from the potentially advantageous properties of chaotic modes, especially because of its spread-spectrum property Deane et al. (1999). Switching power converters are an important source of electromagnetic interference (EMI), so that chaotic regimes could be used as an EMI reduction technique Hamill et al. (1997), Giral et al. (2001), Morel et al. (2004) and Mukherjee et al. (2005).

The control methods of chaos could be classified into different types according to Chen et al. (1993). Although such work classifies the methods for a generic nonlinear system it could be used for the particular case of switching converters: via external force or via feedback control techniques.

1.4.1 Via External Force

The option of adding an external signal to ensure stability has been widely used in the power converter area, especially in CMC, where FSI is exhibited in principle when the duty cycle of the control signal is above 0.5. A common way for avoiding

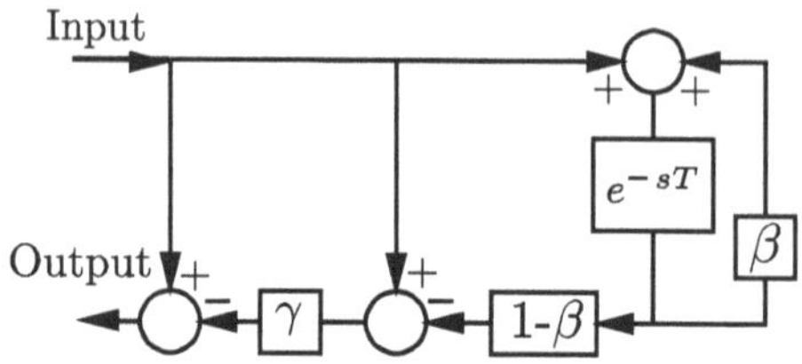

Fig. 1.14 TDFC ($\beta = 0$)and its extension version (ETDFC) schemes

subharmonic oscillations and chaotic regimes is by means of adding a compensating ramp to the feedback control signal Murdock (1987) and Tse and Lai (2000).

Complementarily to adding a ramp, other techniques have been used such as adjusting some parameter of the system, which was initially proposed in Pettini and Lima (1990). Some articles have extended such idea into the power switching converter field, by moving external parameters such as the reference signal. It is the case explored in Zhou et al. (2005) for the case of CMC buck-boost converter. The paper explores the effects of the reference signal parameters (amplitude and phase) upon the FSI boundary.

1.4.2 Via Feedback Techniques

Feedback control is understood as time or frequency based techniques applied by adding a specific circuit to the system feedback loop to ensure the control of instabilities, including chaotic regimes. Whereas these techniques are very common to control SSI due to the availability of design-oriented models allowing a frequency domain representation and a simple stability analysis (*Nyquist* conditions), compensators to control FSI are more based on studying discrete-time models or Lyapunov exponents than in a design-oriented circuit-based understanding.

Well-known techniques are based on Time-delay Feedback Control (TDFC) Pyragas (1992) and Chen and Yu (1999). From such works, the delay-based control techniques have been translated into the power converter area Batlle et al. (1997) and Toribio and Gaston (2001). This controller, also called Time-Delay Auto-Synchronization (TDAS), is shown in Fig. 1.14, uses a control signal conformed with the difference between the current state and the state delayed by one period. The main drawback of this technique is the complex implementation of the delay in the analog domain and the difficulties to understand the effect upon stability boundary depending on the value of its parameters. Note that due to its discrete nature it can very useful for digital application as it has been recently used in Corradini et al. (2008) and Kaouba et al. (2010). Since chaotic behavior can be found in different kinds of systems, work within the field of chaos control has been also considered in other disciplines such as laser stabilization Blakley et al. (2004).

More simple implementation of a chaos control technique has been proposed from a frequency domain in Wei-Guo et al. (2010) but using a notch and a high-pass filter in a buck-based converter. In Athalye and Grantham (1995) and Ahlborn and Parlitz (2006) a simple notch filter is used to control chaos in Duffing's oscillator and laser, respectively. A complementary technique is based on developing more practical (but complex) circuits, such as Poddar et al. (1995) for a CMC boost converter, based on detecting and stabilizing the chaotic attractor by memorizing the last cycle value, which is equivalent to adding a delay module.

Chapter 2
Complex Behavior of VMC Buck Converter: Characterization

The previous chapter has reviewed trends in advanced power management architectures, considering different converters, controllers and functionalities, along with revisiting the available models to predict the dynamics of such architectures.

This chapter is focused on exploring and characterizing the different behaviors that can be exhibited in a switching power converter under a VMC in two ways. On the one hand the interest is centered on exploring the effect of the whole design parameter space parameters upon system dynamics by dividing the multidimensional design parameter space in different stability regions. On the other hand, it is focused on characterizing the dynamic behavior in such regions from specific power-oriented metrics in order to quantify whether or not they are of interest depending on the front-end application.

2.1 Design-Space Characterization of a Voltage-Mode Buck Converter

A switching power system is composed of different nature parameters, encompass those of the converter, controller and modulator.

This section is focused on characterizing the dynamic regimes in different regions of such parametric space starting from a VMC buck converter under a PI control and PWM modulator, shown in Fig. 2.1. The PI controller can be described in the frequency domain as:

$$G_c(s) = k_p \frac{s + \omega_{z1}}{s} \tag{2.1}$$

Considering this system, which is simple but representative of classical voltage-mode regulation, its complete design parameter space W is 9-dimensional:

$$W := (V_g, V_{ref}, L, C, R, k_p, \omega_{z1}, V_m, f_s) \in \mathbb{R}^9 \tag{2.2}$$

E. Rodríguez Vilamitjana et al., *Chaos in Switching Converters for Power Management*,
DOI: 10.1007/978-1-4614-2128-3_2,

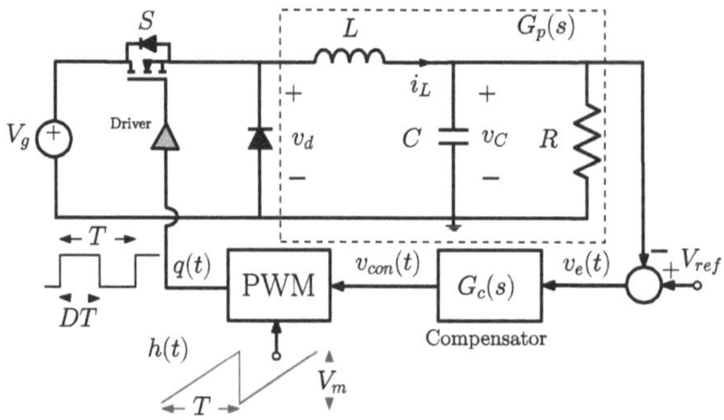

Fig. 2.1 VMC buck converter with a control $G_c(s)$ and PWM modulator controller

Fig. 2.2 Descriptive working regions in CCM

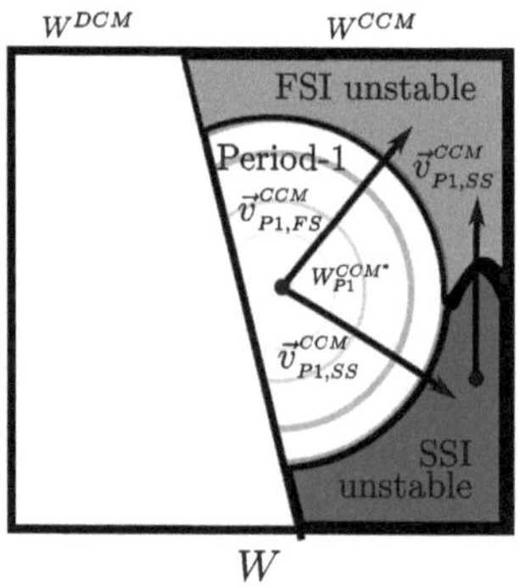

where V_g is the input voltage, V_{ref} is the voltage reference, L is the inductance, C is the output capacitance, R is the output resistance value, k_p is the proportional gain, ω_{z1} is the zero of the PI compensator, V_m is the PWM ramp amplitude and f_s is the PWM ramp frequency (coinciding with the switching frequency in stable regime).

In switching power converters, the design parameter space W can be categorically split into two regions since they can work in two different conduction modes which fundamentally alter the switching process and hence the resulting dynamics. The boundary between both modes is determined by the condition resulting in a null current at the end of the switching period. Boundary between both modes is well known, and, for a buck converter and assuming low-output ripple, it can be expressed as:

$$\frac{\overline{D}}{D} = \frac{2Lf_s}{R} \tag{2.3}$$

where D is the duty cycle of the PWM signal and $\overline{D} = 1 - D$. The next sections are centered on characterizing the effect of the whole design parameter space upon the system dynamics and the boundaries between the different dynamical behavior.

2.1.1 Dynamics of a VMC Buck Converter Working in CCM

In general, the different kinds of dynamics that can be exhibited in a switching power converter can be classified according to their periodicity, as it was mentioned in Sect. 1.3.1: period-one, in which the system periodicity is the switching frequency, FSI, including period-doubling, subharmonic oscillations and chaotic behavior, and SSI, characterized by the exhibition of low-frequency oscillation.

A qualitative map, considering the whole CCM design parameter space W^{CCM} and the possible kinds of behaviors and the related trajectories among them $\mathbf{v}^{\mathrm{CCM}}$, is shown in Fig. 2.2. Note that within the generic FSI and SSI regions, each of them could be subdivided and consist of more complex behaviors, thus being possible to find different dynamics such as period-doubling, period-four, or chaotic behavior in the case of the FSI region.

The numerical characterization in Figs. 2.3, 2.4 and 2.5 show one-dimensional trajectories by sweeping one parameter of the design parameter space W^{CCM}. Such dynamic routes are plotted by means of bifurcation diagrams, which consist of sampling one state variable at the switching frequency which allows to identify the periodicity of the behavior hence classifying it. The parameter values used in this chapter, corresponding to the starting point W^{CCM^*}, are: $V_g = 6\,\mathrm{V}$, $V_{ref} = 3\,\mathrm{V}$, $R = 2.5\,\Omega$, $L = 66\,\mathrm{nH}$, $C = 20\,\mathrm{nF}$, $f_s = 50\,\mathrm{MHz}$, $V_m = 1\,\mathrm{V}$, $k_p = 3$ and $\omega_{z1} = 10\,\mathrm{Mrad/s}$. These values correspond to a miniaturized converter aiming on-chip integration (Villar and Alarcon 2008).

Note that the previous diagrams lead to exhibit different kinds of instabilities depending upon which parameter is swept.

Starting from a stable point in period-one (P1) within the CCM region $W_{\mathrm{P1}}^{\mathrm{CCM}^*}$, it is possible to define different routes depending upon the initial and final states. Observing the previous bifurcation diagrams, we can identify two possible routes: ending up in FSI region $\mathbf{v}_{\mathrm{P1,FS}}^{\mathrm{CCM}}$ or ending up within SSI region $\mathbf{v}_{\mathrm{P1,SS}}^{\mathrm{CCM}}$.

According to the previous nomenclature and with the previous simulation results, the routes starting from a stable period-one behavior within CCM region $W_{\mathrm{P1}}^{\mathrm{CCM}}$ can be split as:

$$\mathbf{v}_{\mathrm{P1,FS}}^{\mathrm{CCM}}(L, C, f_s, V_g, V_{ref}, k_p)$$
$$\mathbf{v}_{\mathrm{P1,SS}}^{\mathrm{CCM}}(R, C, \omega_{p1})$$

The routes shown in the bifurcation diagrams are only valid for the given starting point $W_{\mathrm{P1}}^{\mathrm{CCM}^*}$ and moving along one dimension, and hence they can not be considered univocal for the other points within W^{CCM}. Therefore, it is considered an alternative exploration of such design parameter space in order to determine the possible routes. The exploration of the design parameter space pair (k_p, ω_{z1}), which is shown in Fig. 2.6, unveils complex boundaries dependency of both parameter values.

Further exploration shows that other parameters can lead to both kinds of behavior depending upon the starting point. For instance, this occurs in the case of proportional

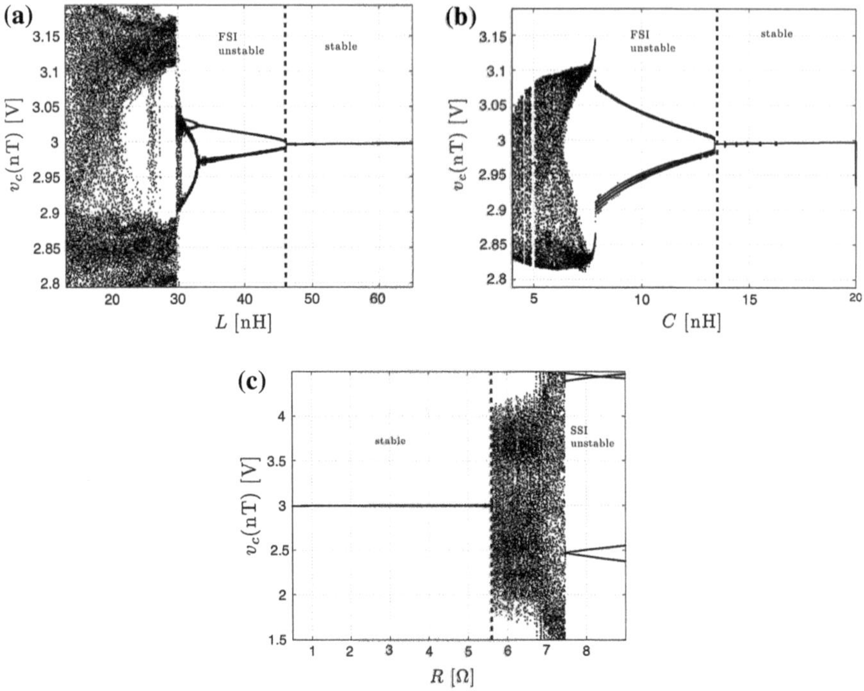

Fig. 2.3 Bifurcation diagrams obtained by sweeping different converter parameters. **a–b** FSI, **c** SSI

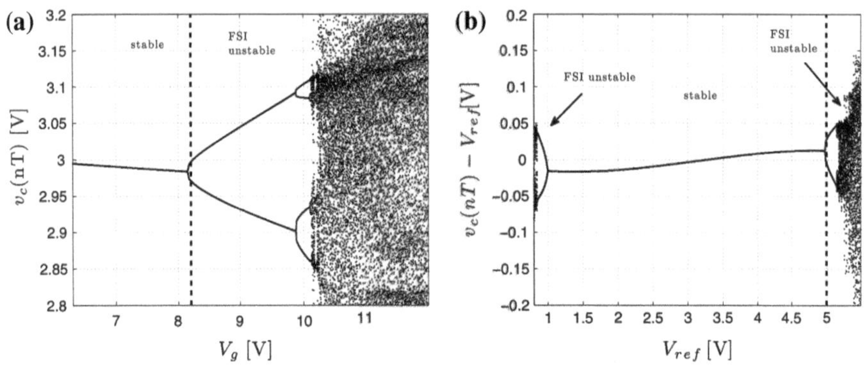

Fig. 2.4 Bifurcation diagrams obtained by sweeping the input and the reference voltages V_g and V_{ref} respectively

gain k_p or modulation amplitude V_m which can lead to either SSI or FSI depending on other parameters of the system such as R, C or ω_{z1}.

While the previous routes were only carried out in a one-dimensional space, limiting the exploration of the dynamics, the definition of more complex routes,

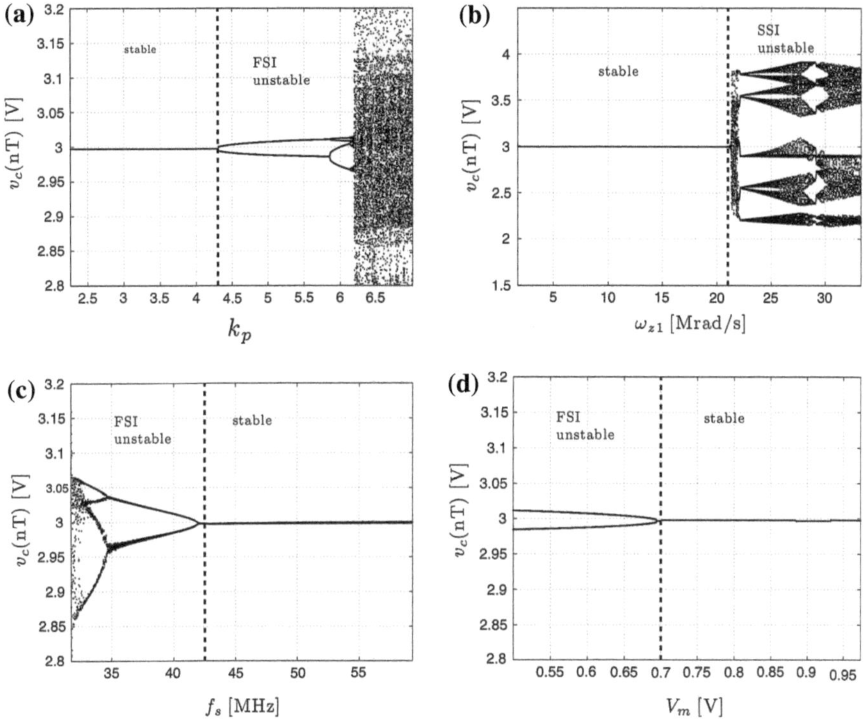

Fig. 2.5 Bifurcation diagrams obtained by sweeping controller parameters. **a** Proportional gain k_p. **b** PI zero ω_{z1} and modulator parameters. **c** Switching frequency f_s and **d** ramp amplitude V_m

depending on multiple swept parameters, allows characterizing the whole dynamic behavior in a more flexible way, such as the transition between different dynamics regimes and stability regions, which by exploring as a function of only one parameter can not be reached.

Therefore, given the n-dimensional design parameter space W, composed of n parameters p_i (for $i = 1$ to n), it is possible to define a route $\mathbf{v_y}(y)$ depending on only one variable (y), which compresses the variation of other parameters p_i into one route:

$$p_i = f_i(y) \quad \text{for } i = 1 \text{ to } n \tag{2.4}$$

$$\mathbf{v}_x(p_1, p_2, p_3, \ldots, p_n) = M \rightarrow \mathbf{v_y}(y) = \mathcal{P} \tag{2.5}$$

Note that different routes will be defined as a function of the parameter $\mathcal{P}$, namely constant level variable.

As a simple example, a linear route from SSI to FSI $\mathbf{v}_{\text{FS,SS}}^{\text{CCM}}$, shown in Fig. 2.6, is defined over the parametric space by the pair proportional gain-compensator zero (k_p, ω_{z1}), with starting point (3, 25 Mrad/s) and end point (5, 10 Mrad/s). Therefore,

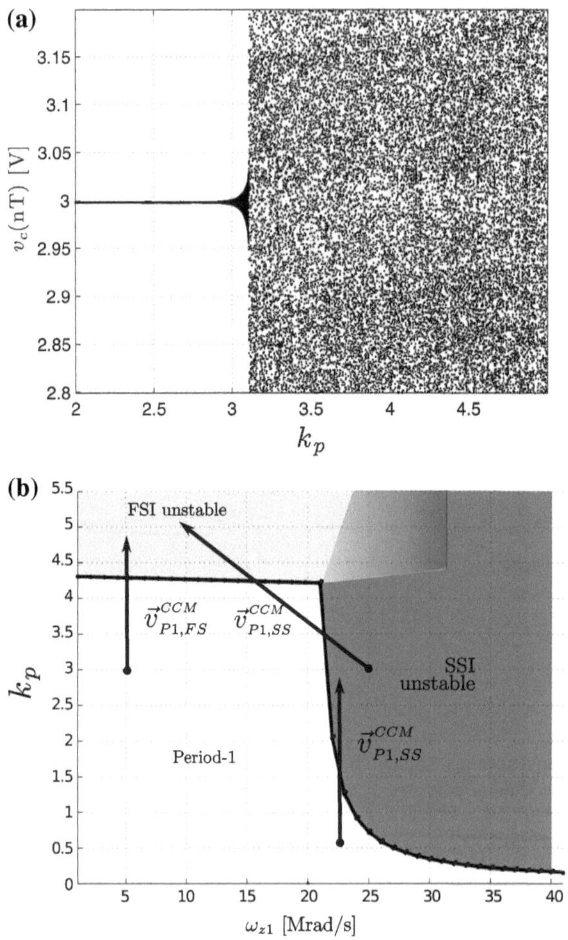

Fig. 2.6 **a** Bifurcation diagram obtained by sweeping proportional gain k_p with $\omega_{z1} = 21$ Mrad/s. **b** Stability boundary characterization over the space (k_p, ω_{z1}) along with possible routes

under such condition, the route can be expressed as:

$$\mathbf{v}^{\mathrm{CCM}}_{\mathrm{FS,SS}} = f(k_p, w_{z1}) = \mathcal{P} \rightarrow k_p + \frac{2}{15}(\omega_{z1}10^{-6} - 10) = \mathcal{P} \tag{2.6}$$

The route $\mathbf{v}^{\mathrm{CCM}}_{\mathrm{FS,SS}}(k_p, \omega_{z1})$ intends to explore the transition from FSI to SSI regions. Finally, the routes can be expressed only as a function of such variable y:

$$f(k_p, \omega_{z1}, P) \equiv f(y, \mathcal{P}) \tag{2.7}$$

in which each parameter can be expressed as a function of variable y:

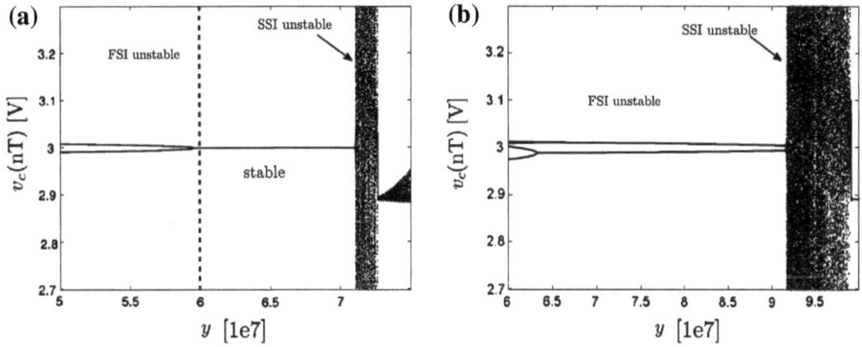

Fig. 2.7 Route $\mathbf{v}^{CCM}_{FS,SS}$ by plotting the bifurcation diagram as a function of y. **a** $\mathcal{P} = 5$ and **b** $\mathcal{P} = 6$

$$\omega_{z1}(y) = y10^6 \tag{2.8}$$

$$k_p(y, \mathcal{P}) = \mathcal{P} - \frac{2}{15}(y - 10) \tag{2.9}$$

The route as a function of the variable y is shown in Fig. 2.7, obtaining a bifurcation diagram that characterizes the transition between FSI and SSI regions. Note that the boundary is abrupt, without coexistence of attractors.

2.1.2 Dynamics in VMC Buck Converter Operating in DCM

The next characterizations explore the effect of DCM upon the dynamic behavior, especially centered on the effect of crossing the boundary between both modes of the system. The parameters values used in this section are $V_g = 6$ V, $V_{ref} = 3$ V, $R = 6\,\Omega$, $L = 45$ nH, $C = 50$ nF, $f_s = 25$ MHz, $V_m = 1$ V, $k_p = 2.5$ and $\omega_{z1} = 1$ Mrad/s.

It is worth observing that having thoroughly explored the design parameter space in DCM, it has not been identified the exhibition of SSI. Then, considering only the existence of two behaviors in DCM, namely period-one and FSI, along with the abrupt nature of DCM boundary, the set of routes within the whole design parameter space and the dynamic analysis complexity is increased, as it is shown in the working regions in Fig. 2.8, depending on which parameters are swept.

Considering the boundary expression between the operating modes, shown in Eq. (2.3), the route from CCM to DCM is:

$$\mathbf{v}^{CCM,DCM}(L, f_s, R, D) \tag{2.10}$$

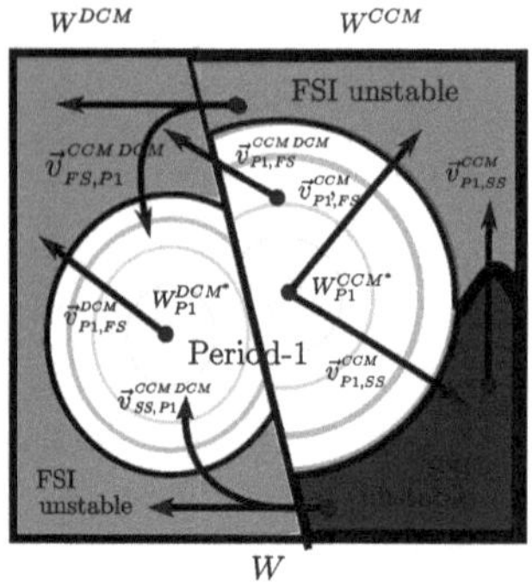

Fig. 2.8 Descriptive working regions in DCM and CCM along with routes between modes and behaviors

The numerical simulation in Fig. 2.9 shows the effect of crossing from CCM to DCM considering different routes, obtained by sweeping different parameters and starting from different CCM conditions.

Different behaviors can be observed depending on which parameter is swept. The decrease of switching frequency may turn the converter into DCM and, once in DCM, it can lead to exhibit FSI in a similar way as it occurs in CCM. Regarding the inductance, the effect is opposite to its effect in CCM, since, as it is decreased, it cancels the period-doubling behavior once in DCM. Furthermore the effect of the load resistance on the SSI bifurcation is similar to the inductance since it also ends up canceling the instability once in DCM, firstly leading to FSI and subsequently leading to period-one behavior if it is decreased. Finally, it can also occur that crossing the conduction boundary from period-one in CCM leads to an abrupt jump into FSI, as it is shown by sweeping the voltage reference V_{ref}.

The effect of other parameters, which do not imply a change into DCM, such as the proportional gain k_p, are shown in Fig. 2.10 showing a similar effect as in the CCM case.

To summarize, we can observe that a VMC buck converter considering both conduction modes can be very complex, especially when the DCM is considered. This section has demonstrated that the different behaviors depend upon different parameters and that the same parameter can lead to different bifurcation scenarios depending upon the combination with other parameters of the multi-dimensional design parameter space. Furthermore, such complexity is increased when the route crosses the conduction mode boundary, which can lead to an abrupt transition in dynamics. While hitherto a qualitative characterization of the system behavior has been carried out focusing on pointing out the rich dynamic behavior of the switching power converter and exploring the effect of different nature parameters within the design parameter space, the next section is focused on a quantitative characterization of such dynamic behavior by establishing a set of power-oriented metrics that allow relating converter behavior properties to power application requirements for a given dynamic regime encountered in the design parameter space.

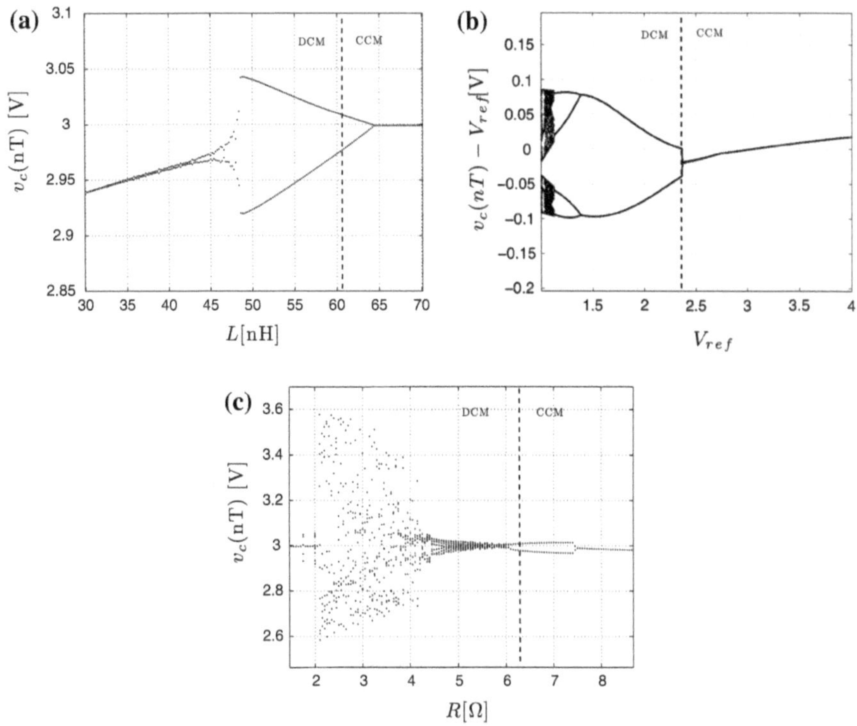

Fig. 2.9 Different bifurcation diagrams obtained by crossing from CCM to DCM. **a** $\mathbf{v}_{\text{FS,P1}}^{\text{CCM,DCM}}(f_s)$ as a function of switching frequency f_s. **b** $\mathbf{v}_{\text{FS,P1}}^{\text{CCM,DCM}}(L)$ as a function of inductance L. **c** $\mathbf{v}_{\text{P1,FS}}^{\text{CCM,DCM}}(V_{ref})$ as a function of voltage reference V_{ref} with $k_p = 2$. **d** $\mathbf{v}_{\text{SS,P1}}^{\text{CCM,DCM}}(R)$ as a function of output resistor R and with $\omega_{z1} = 10\,\text{Mrad/s}$

2.2 Power-Oriented Electrical Metrics Characterization of VMC Buck Converter Dynamics

This section is focused on a quantitative characterization of the instabilities that can be exhibited by a switching power converter, unveiling the impact they can have upon the system performances, with special attention on power-oriented electrical metrics, such as ripple, switching frequency or spectral behavior, which affect these performance.

Most of the previous works regarding the study of stability boundaries have been focused on predicting the exhibition of the different dynamic behaviors, but usually ignore evaluating the impact of such behaviors upon the system performance taking into account their final application. In such works, instabilities are analyzed from a qualitative standpoint, hence representing the bifurcation diagram or the state-plane, in which it is possible to observe the evolution of the state variables.

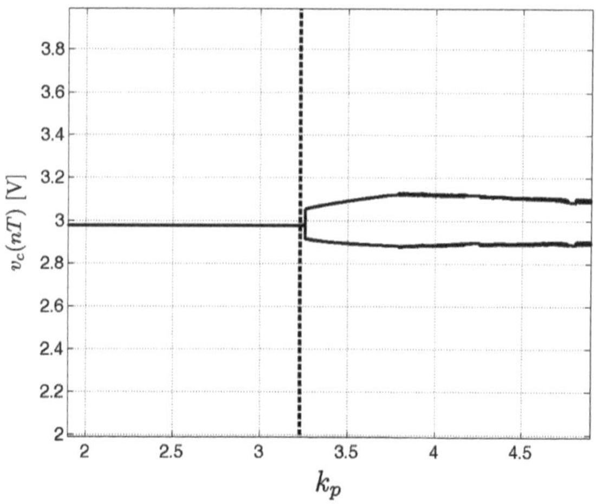

Fig. 2.10 Bifurcation diagram by sweeping the proportional gain k_p

A complementary approach from an engineering standpoint, has been initially proposed in some works focusing on the effect of such instabilities in the spectrum (Banerjee et al. 2002; Deane and Hamill 1996; Giral et al. 2001; Woywode et al. 2003), in order to reduce electromagnetic interference of switching power converters. The idea is based on taking advantage of the well-known spread spectrum feature of chaotic behavior to reduce the harmonics at high frequencies.

However, the work presented in this section, addresses the characterization of the different instabilities not only on spectral performance, but also taking into account other power-oriented metrics, including output ripple and averaged switching frequency.

The fact that chaotic behavior can be exhibited requires to use a statistical tools, in a similar way as it has been done in (Woywode et al. 2003) in order to properly characterize the system. Therefore the chosen metrics are:

- The standard deviation of the output voltage waveform σ_{v_c}, in a steady-state period of time, as a measure of output ripple. Note that considering the average value 0, the RMS value coincides with the variance of the signal.
- The averaged switching frequency $\langle f_s \rangle$, understood as the average number of switching events per ramp period $T = 1/f_s$.

Note that while the previous characterization, in spectrum or state-variable is based on taking a photography under a given dynamic regime, the availability of such metrics allow to build an extended bifurcation diagram, showing the evolution of each metric by sweeping a given parameter of the parameter space.

The first exploration attempts to distinguish and compare the qualitative difference between both SSI and FSI. This is shown in Figs. 2.11 and 2.12 respectively.

In Fig. 2.11 the bifurcation diagram is shown by sweeping the proportional gain k_p, which ends up in chaotic behavior. The bifurcation diagram is complemented with the evolution plot of the averaged switching frequency $\langle f_s \rangle$ and standard deviation

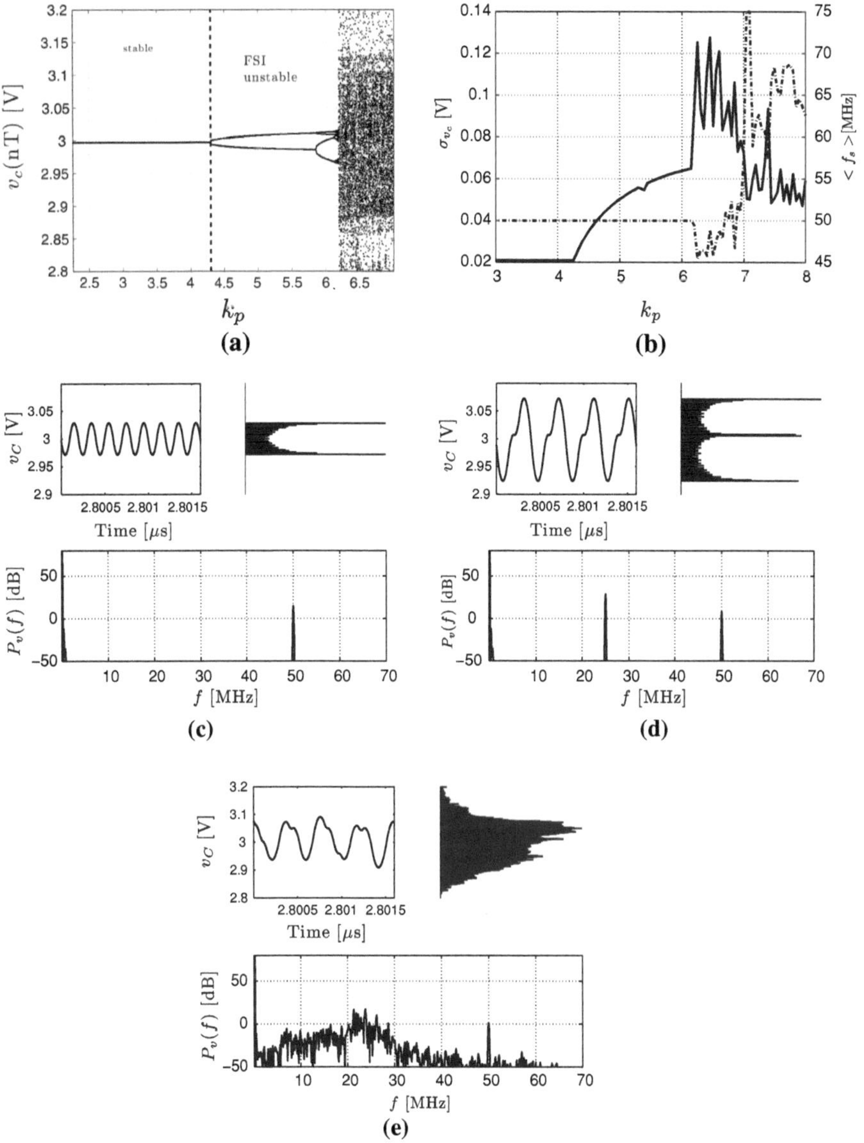

Fig. 2.11 **a** Bifurcation diagram as function of k_p. **b** Standard deviation σ_{v_c} (*solid*) and averaged switching frequency $\langle f_s \rangle$ (*dashed*). Representative time-domain waveform, histogram and spectrum (calculated within 1,000 periods) of the output voltage for **c** $k_p = 3$ (period-one). **d** $k_p = 5$ (period-doubling) and **e** $k_p = 7$ (chaotic regime) with $D = 0.5$

of output voltage σ_{v_c}. The exploration shows that when period-doubling is exhibited, the ripple steadily increases. Once into chaotic regime the ripple starts decreasing, but on the other hand the averaged switching increases. This can be understood because

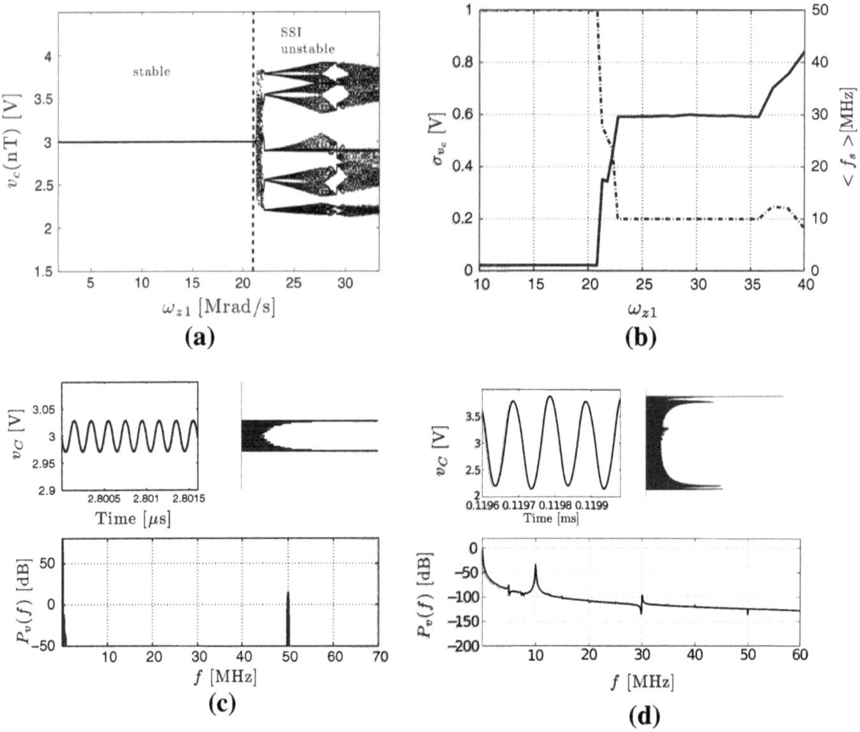

Fig. 2.12 **a** Bifurcation diagram as function of ω_{z1}. **b** Standard deviation σ_{v_c} (*solid*) and average switching frequency (*dashed*) $\langle f_s \rangle$. Representative time-domain waveform, histogram and spectrum (calculated within 1,000 periods) of the output voltage for **c** $\omega_{z1} = 10$ Mrad/s (period-one) and **d** $\omega_{z1} = 25$ Mrad/s (SSI) with $D = 0.5$

of the exhibition of other instabilities (border collision) such as sliding behavior that entails multiple crossing of the ramp signal in one PWM ramp period. The spectrum of the different stages of the route to chaos entails the exhibition of subharmonics and, once into chaotic behavior, the well-known spread spectrum effect.

On the other hand, the effect of SSI into power metrics is different as it is possible to observe in Fig. 2.12. The exhibition of SSI leads to high reduction of the switching frequency and then a high increase in terms of ripple.

In addition, in Fig. 2.13 it is shown that the bifurcation diagram can vary depending upon the other parameters of the converter such as the duty cycle, which modifies the characteristic route of chaos: it is still composed of period-doubling exhibition, but it has different power metrics. Numerical simulations show that the averaged switching frequency is kept under the ramp frequency once within the chaotic regime and the ripple tends to decrease. This can be explained by the fact that being the duty cycle different of 0.5, the voltage applied to the inductor is unbalanced (in average) and this causes the system to exhibit skipping cycle phenomena hence reducing the number switching per cycle.

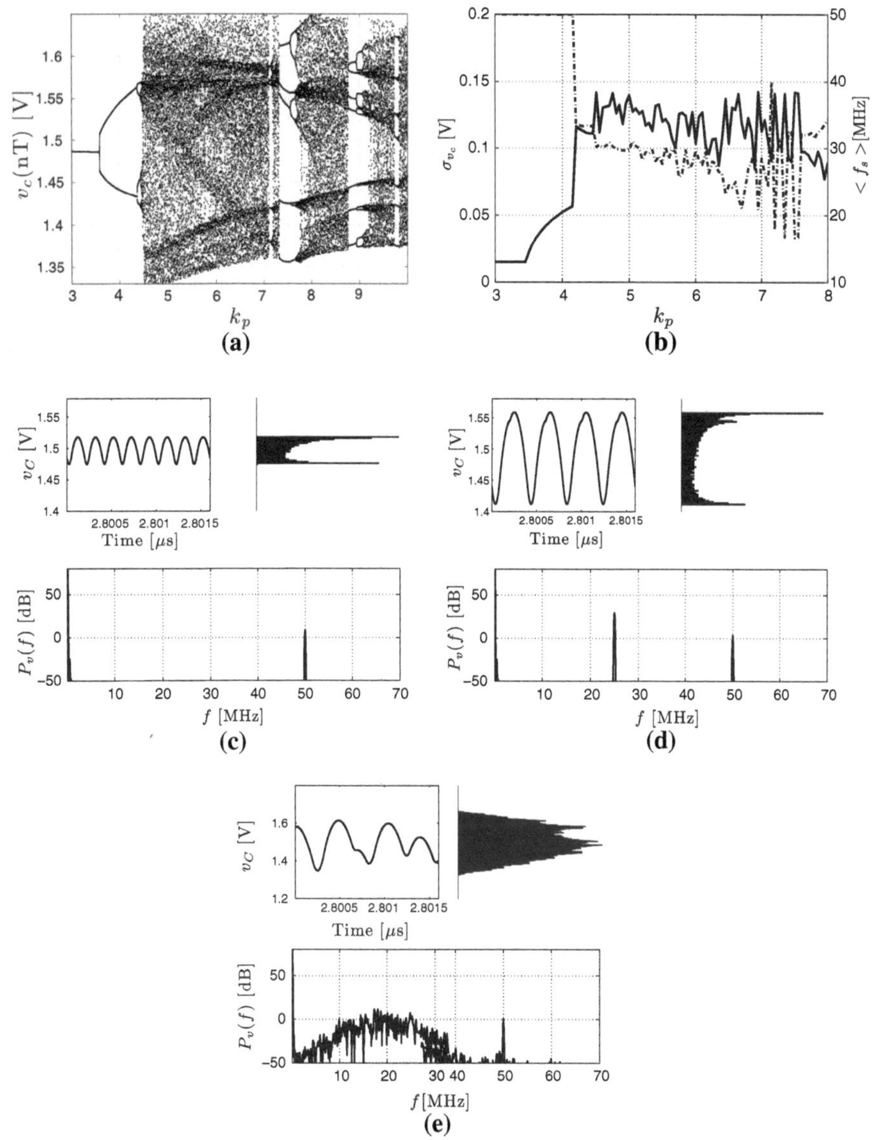

Fig. 2.13 **a** Bifurcation diagram as a function of k_p. **b** Standard deviation σ_{v_c} (*solid*) and average switching frequency $\langle f_s \rangle$ (*dashed*). Representative time-domain waveform, histogram and spectrum (calculated within 1,000 periods) of the output voltage for **c** $k_p = 3$ (period-one). **d** $k_p = 4$ (period-doubling) and **e** $k_p = 7$ (Chaotic behavior) with $D = 0.25$

2.3 Concluding Remarks

This chapter has focused on characterizing the different dynamic behaviors that a VMC buck converter with PI controller can exhibit. First, the chapter has identified the multidimensional design parameter space and characterized the dependencies of the stability boundaries upon the design parameter space. Numerical simulations based on the exact instantaneous state equations have shown rich and complex dynamic phenomena depending upon which parameter is swept. Furthermore, the chapter has discussed that trends towards miniaturization, namely reducing reactive components value or the switching frequency, can make the converter more prone to exhibit FSI, hence validating the interest on being able to predict such instabilities from a design-oriented standpoint and to synthesize improved dynamic controller. Finally, the chapter has quantified how the different dynamic regimes affect the switching converter from a power management perspective, by characterizing different power processing performance metrics such as the output voltage ripple (related to the quality of the DC–DC conversion), spectral behavior (related to EMI aspects) and average switching frequency (related to efficiency through the dominant switching losses).

Chapter 3
Design-Oriented Models for Predicting Instabilities in a Buck Switching Power Converter

The previous chapter has identified and have explored the fact that trends towards integration and miniaturization of switching power converters, such as reducing reactive component parameters or the switching frequency, can lead the system to exhibit FSI, as it has also been previously observed in (Villar et al., 2002) and (Allard et al., 2004).

This fact reactivates a renewed interest in the need of analytical tools for predicting such instabilities, especially for VMC, for which there is a lack of design-oriented stability models despite its widespread use in regulation applications. Beyond the fact that there has been a considerable effort in characterizing the FSI stability boundary by means of discrete-time models (Fossas and Olivar, 1996), (di Bernardo et al., 1997), these models only allow a parametric characterization of the stability boundary neglecting the system-circuit interpretation. In parallel, the existing models derived from a circuit interpretation, such as the averaged models, can not predict FSI due to their averaging nature hence lacking of the switching information, as it will be shown at the beginning of this chapter.

A ripple-based stability condition for a buck converter is proposed in this chapter. This index captures the effect of different nature converter parameters upon the FSI boundary, thereby allowing a compact design-oriented circuit-based stability criterion and facilitating further investigation in enhanced chaos controllers oriented to avoid these instabilities.

First, the averaged model -with its benefits in modeling the system from a design-oriented circuit-based standpoint-, and the discrete-time model,-with its capability to predict the overall stability are compared.

Subsequently, having demonstrated the inability of the averaged mode for predicting FSI, an alternative ripple-based index is presented for predicting FSI from a circuit-based standpoint. First, the approach is validated for a VMC buck converter in a wide design parameter space by means of numerical simulation of the exact state equations, through experimental measurements and comparing both with the results obtained from the discrete-time model. Furthermore, it is demonstrated that the ripple index can also be derived from such discrete-time model, hence giving

E. Rodríguez Vilamitjana et al., *Chaos in Switching Converters for Power Management*,
DOI: 10.1007/978-1-4614-2128-3_3, © Springer Science+Business Media New York 2013

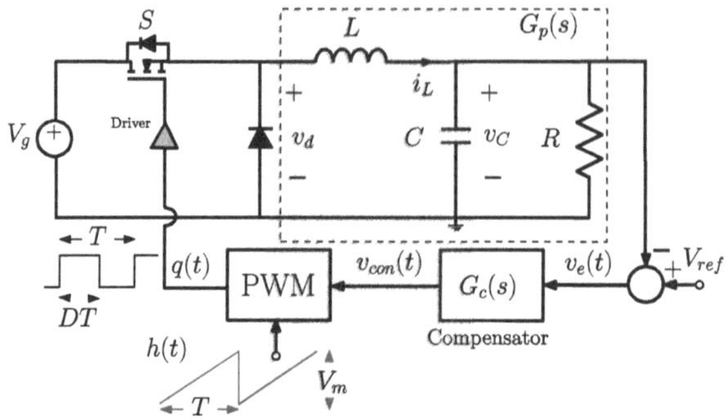

Fig. 3.1 Circuit diagram of a DC-DC VMC buck converter with a PWM

mathematical support to extend the approach to other controls. The novel stability index has been combined along with the averaged model leading to a full characterization of the stability boundaries of the system. Subsequently, the ripple-based index is also extended to the DCM case and to the CMC buck converter, further proving the benefits of the approach and its general purpose applicability. Finally, the chapter concludes by providing a frequency domain model, based on building a new modulator transfer function, capable to predict the overall stability margin without losing the simplicity/circuit standpoint, thereby also facilitating the derivation of new controllers, that will be tackled in further chapters.

3.1 Design-Oriented Averaged Model: Benefits and Limitations

Averaged models can be obtained starting from a circuit bottom-up approach in which the circuit state variables, inductor current and capacitor voltage, are averaged during one switching period and then linearized (small-signal model) with respect to the system external variables such as the input voltage or driving signal duty cycle (Middlebrook and Cuk, 1976).

The closed-loop stability analysis of the whole system can be carried out from the total loop transfer function $T_{avg}(s)$ which requires to take into account the converter control-to-feedback and modulator small-signal transfer functions, $H_{d,avg}(s)$ and $H_{m,avg}(s)$ respectively, along with the compensator frequency response $G_c(s)$:

$$T_{avg}(s) = H_{d,avg}(s)G_c(s)H_{m,avg}(s) \tag{3.1}$$

For the particular case of a VMC buck converter case under a PI compensator and constant switching frequency PWM modulator, shown in Fig. 3.1, these transfer functions are given by:

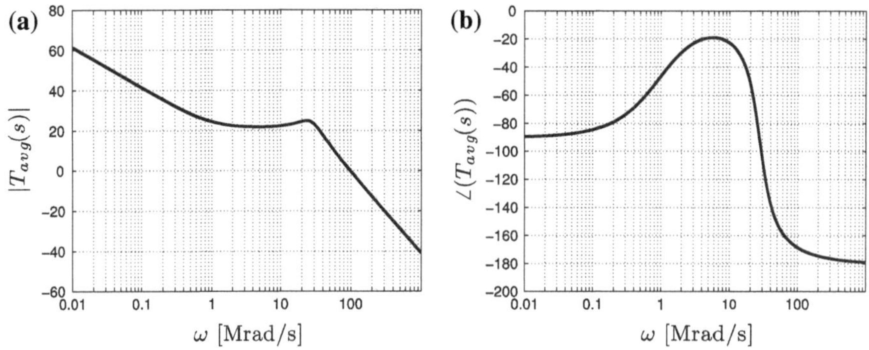

Fig. 3.2 Bode plot of total loop gain $T_{avg}(s)$ of a buck converter under PI compensator. **a** magnitude and **b** phase

$$H_{d,avg}(s) = V_g \frac{\omega_0^2}{s^2 + \omega_{RC}s + \omega_0^2} \tag{3.2}$$

$$G_c(s) = k_p \frac{s + \omega_{z1}}{s} \tag{3.3}$$

$$H_{m,avg}(s) = \frac{1}{V_m} \tag{3.4}$$

where $\omega_0 = 1/\sqrt{LC}$ and $\omega_{RC} = 1/(RC)$ are supposed to be smaller than $\omega_s = 2\pi f_s$ the angular switching frequency of the ramp modulator.

The averaged model is naturally expressed in the s-domain, which allows a frequency domain representation, such as the Bode diagram shown in Fig. 3.2 which consists of independently plotting the magnitude and phase of $T_{avg}(s)$, hence facilitating the stability analysis by means of the Nyquist stability conditions:

$$\begin{aligned} &\angle T_{avg}(j\omega_p) > -180^\circ \text{when } |T_{avg}(j\omega_p)| = 1 \\ &|T_{avg}(j\omega_g)| < 1 \text{when } \angle T_{avg}(j\omega_g) = -180^\circ \end{aligned} \tag{3.5}$$

The Nyquist stability conditions lead to establish two metrics to quantify the propensity to exhibit instabilities, namely phase margin and gain margin. The first one is the phase difference to -180° when the gain is 0 dB. The second, complementary to the first one, is the loop gain magnitude respect to 0 dB when the phase reaches -180°.

Furthermore, apart from such fundamental stability analysis, the frequency representation facilitates a clear design standpoint hence allowing to independently address the synthesis of the transfer function of each system block from a pole-zero interpretation so as to properly adjust them to the system stability and dynamics requirements.

Despite all these benefits, due to the averaged nature of the model, it has an important limitation in predicting the complete instability phenomena, since it is

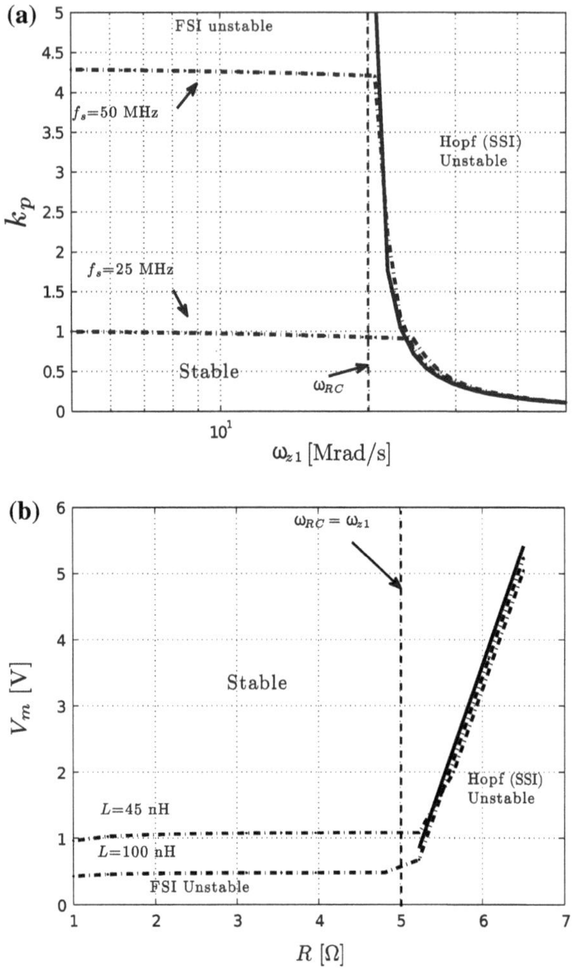

Fig. 3.3 Overall stability boundary map in the design parameter space **a** ω_{z1}, k_p and f_s (b) V_m, R and L obtained from the discrete-time mode (*dash-dot*) and results from the averaged model using the Nyquist stability criterion (*solid*)

not able to predict FSI boundary as it is shown in Fig. 3.3 where results from such model are compared to the ones obtained from the discrete-time model, developed in Appendix A. The figures show the accuracy of the averaged model for predicting the SSI boundary but also its incapability to predict the FSI boundary.

The value of the parameters used for Fig. 3.3 are $V_g = 6\,\mathrm{V}$, $V_{ref} = 3\,\mathrm{V}$, $R = 2.5\,\Omega$, $L = 66\,\mathrm{nH}$, $C = 20\,\mathrm{nF}$, $f_s = 50\,\mathrm{MHz}$, $V_m = 1\,\mathrm{V}$, $\omega_{z1} = 10\,\mathrm{Mrad/s}$ and $k_p = 3$. As in the previous chapter, these values correspond to a miniaturized converter aiming on-chip integration (Villar and Alarcon, 2008), but the selection is representative - through scaling- of any converter exhibiting moderately large ripple ($\sim$50 mV).

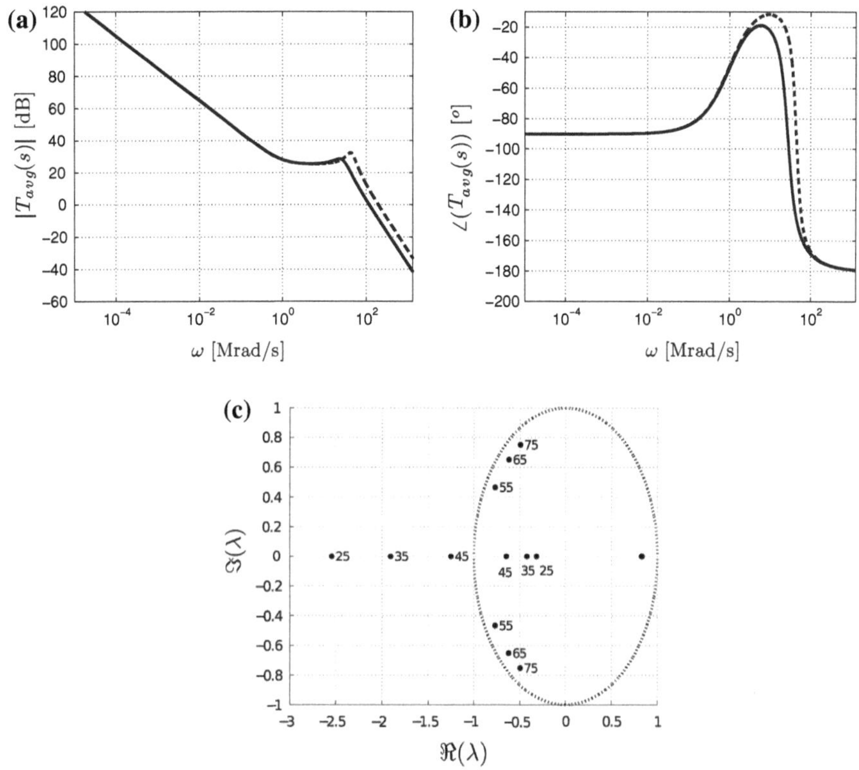

Fig. 3.4 Bode plot by reducing the inductance L: $L = 66\,\text{nH}$ (*solid*) $L = 25\,\text{nH}$ (*dashed*) **c** Eigenvalues evolution from the system discrete-time model by sweeping the inductance (*values are in nH*)

Despite the aforementioned clear benefits of the frequency representation, it can be difficult to identify the effect of the system parameters upon stability boundaries.

Let us study the effect of the inductance upon stability due to its interest for integration. From the Bode plot, shown in Fig. 3.4, it can be observed that by reducing the inductance value, it has an influence on both magnitude and phase, hence being difficult to know its effect upon stability boundary. An alternative representation of the system stability is by means of plotting the eigenvalues of the Jacobian matrix of the discrete-time model of the system and checking their position with respect to the unit circle. By exploring the loci of the eigenvalues, it is possible to observe that the inductance only affects the phase of the eigenvalues but not their magnitude when these eigenvalues have an imaginary part, hence it can only lead to FSI exhibition, but not to SSI exhibition (recall that SSI are exhibited when unit circle is crossed with imaginary eigenvalues). While from the discrete-time model it can be easily anticipated that reducing the inductance will never lead to SSI exhibition, with the averaged model it is more difficult to deduce such information.

In order to know the effect of each parameter upon stability, a complementary representation is proposed starting from the s-domain loop gain $T_{avg}(s)$ given in Eq. 3.1 and the Nyquist stability criteria (Eq. 3.5). For a VMC buck converter under a PI controller, at the frequency ω_g where $\angle T(j\omega_g) = -180$, one has

$$|T(j\omega_g)| = \frac{k_p}{V_m}\frac{V_g}{1 - LC\omega_g^2} < 1 \tag{3.6}$$

$$\omega_g = \sqrt{\frac{1}{LC - \frac{L}{R\omega_{z1}}}} \tag{3.7}$$

Finally, a closed-form index expression for predicting SSI stability boundary can be expressed, in which it can be observed implicitly which parameters can lead to the occurrence of such instabilities. The SSI index is given by:

$$\rho_{SS} := \frac{k_p}{V_m} V_g \left(\frac{\omega_{z1}}{\omega_{RC}} - 1\right) \tag{3.8}$$

and the stability condition is $\rho_{SS} < 1$. The index-based stability condition is simpler than the Bode diagram and does not provide any information about the frequency domain response of the system or its dynamic behavior but it, implicitly, gives the impact of each parameter upon the stability boundary. Then, from Eq. 3.8, it is possible to observe that the SSI would only occur provided that $\omega_{z1} > \omega_{RC}$, and, in that case, parameters such as k_p or V_g would have the same impact on stability boundary. Note that there is no effect of the inductance on such stability boundary, which is aligned with the discrete-time model results.

The simplicity of the analysis in both cases, through the Bode representation or though the stability index ρ_{SS}, facilitate the circuit designer to understand the effect of each system parameter, thereby allowing to choose the proper parameter values to guarantee the stability or certain dynamics requirements.

However, there is still an open challenge regarding the prediction of FSI using a design-oriented standpoint, especially in VMC converters. Aligned with that aim, an approach has been proposed in (Fang and Abed, 2001), based on a Fourier analysis of both period-one and period-two waveforms, in which a closed-form condition for stability is obtained, but unfortunately this condition is still away from a simple design-oriented expression.

The next section proposes a design-oriented ripple-based index which can capture the effect of different parameters of the converter upon the FSI boundary, as in the case of SSI index given in Eq. 3.8, allowing a circuit-level standpoint prediction and paving the way to further investigation in more adequate controllers oriented to avoid these instabilities.

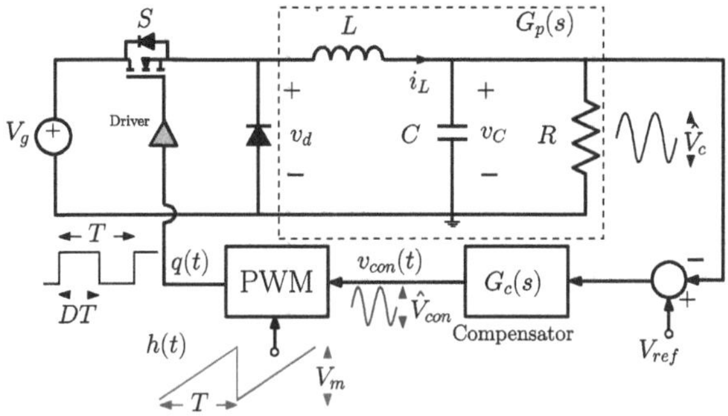

Fig. 3.5 Circuit diagram of a VMC DC-DC buck converter with a compensator frequency response $G_c(s)$ and under a fixed frequency PWM strategy

3.2 Ripple-Based Design-Oriented Index: Hypothesis for Predicting FSI

With the aim of deriving a design-oriented model to predict FSI, a ripple-based index is proposed for VMC converters, as a complementary tool to the average model. This index ρ is defined as the scaled ripple amplitude $\hat{V}_{con}$ of the control signal $v_{con}(t)$ with respect to the modulator amplitude V_m, therefore compressing the different nature converter design parameters, such as the parameters of the reactive components, control coefficients and modulator paramters into a single parameter given by

$$\rho := \frac{\hat{V}_{con}}{V_m} = \frac{\mathcal{F}\hat{V}_C}{V_m} \tag{3.9}$$

where $\mathcal{F}$ models the effect of the controller upon the ripple amplitude $\hat{V}_C$ of the output capacitor voltage v_C. The ripple index hypothesis is based on the observation that when the ripple instability index (Eq. 3.9) reaches a critical value ρ_{crit}, FSI is exhibited. Then the stability condition is:

$$\rho < \rho_{crit} \tag{3.10}$$

The next section validates this ripple-based index hypothesis for the VMC buck converter under a simple PI compensator.

3.2.1 *Validation of the Ripple-Based Approach by Means of the Switched Model*

The most widespread configuration, namely a buck converter with PI feedback compensator, shown in Fig. 3.5, has been chosen to illustrate the validity of the ripple-based approach. Let k_p be the proportional gain of the controller and ω_{z1} its real zero. The parameter ω_{z1} is located at low frequency ($\omega_{z1} < \omega_c$), in order not to induce SSI (see Eq. 3.8). Hence, the effect of the controller upon the converter output voltage ripple is constant ($\mathcal{F} = k_p$) for all frequencies larger than ω_{z1}. This choice of the dynamic controller parameters will allow to independently analyze the effect of the proportional gain k_p and the average value of the control signal on stability.

Although different expressions for the amplitude of the output voltage ripple may be used, the following simple expression is chosen in order to keep the design-oriented standpoint Erickson and Maksimovič (2001):

$$\hat{V}_c := \frac{V_g D\overline{D}}{8LCf_s^2} \tag{3.11}$$

where D is the steady state duty cycle of the PWM driving signal $q(t)$ (See Fig. 3.5) and $\overline{D} = 1 - D$. Accordingly, ρ given in Eq. 3.9 can be expressed as a function of design parameters as:

$$\rho(V_g, L, C, k_p, f_s, D, V_m) := \frac{k_p}{V_m}\frac{V_g D\overline{D}}{8LCf_s^2} \tag{3.12}$$

Figure 3.6 shows different bifurcation diagrams, obtained from the exact switched state equations, by considering parameters of different nature as bifurcation parameters. In all cases, an exhibition of FSI occurs, namely period-doubling cascade which ends, eventually, to chaotic behavior. The value of the fixed circuit parameters used for Fig. 3.6 are $V_g = 6\,\text{V}$, $V_{ref} = 3\,\text{V}$, $R = 2.5\,\Omega$, $L = 66\,\text{nH}$, $C = 20\,\text{nF}$, $f_s = 50\,\text{MHz}$, $V_m = 1\,\text{V}$, $k_p = 3$ and $\omega_{z1} = 1\,\text{Mrad/s}$.

It can be observed in Fig. 3.6 that by representing the bifurcation diagrams in terms of the corresponding ripple-based index ρ defined in Eq. 3.12, the bifurcation boundary remains practically constant (dashed line in Fig. 3.6d–f), independently of the swept parameter, being the FSI critical value $\rho_{crit} \approx 0.245$ in all figures. Further validation of the ripple-based index approach is carried out in Fig. 3.7 for a wide design parameter space. This figure shows the stability boundary obtained from numerical simulations of the circuit state equations contrasted with those obtained from the ripple based index. Although the error increases for low values of the load resistance R, the relatively low error obtained for the rest of the design parameter space validates this design-oriented approach for predicting FSI. The approach is only considered for the system working in the CCM, then the design parameter space is limited by the DCM condition. Further extension to DCM will be tackled later.

Having explored thoroughly the design parameter space, it is obtained that ρ_{crit} is constant, save as dependence upon the duty cycle D. The evolution of ρ_{crit} in terms of the duty cycle is shown in Fig. 3.8. Therefore, the stability condition to avoid FSI exhibition, in terms of the system parameters, can finally be expressed as:

$$\rho(V_g, D, L, C, k_p, f_s, V_m) < \rho_{crit}(D) \tag{3.13}$$

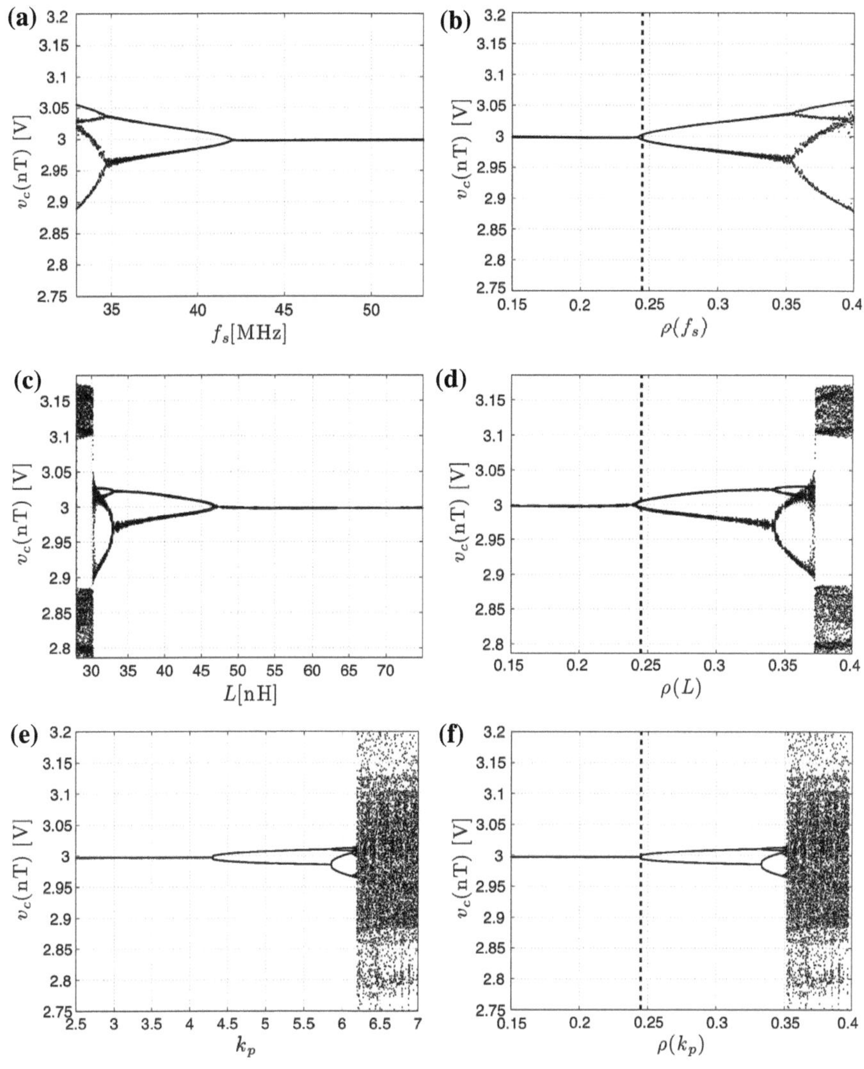

Fig. 3.6 Bifurcation diagrams obtained by sweeping different nature parameters **a** the switching frequency f_s, **b** the inductance L and **c** the proportional gain k_p and its equivalent representation as a function of the ripple-based index ρ **d**, **e** and **f**, respectively. $\rho_{crit} \approx 0.245$ (*dashed line*)

This stability condition given in Eq. 3.13 shows the benefit of the ripple-based approach, which allows compressing most of the design-space parameters into a single ripple-based index ρ, providing a design-oriented tool for predicting the effect of each parameter upon stability.

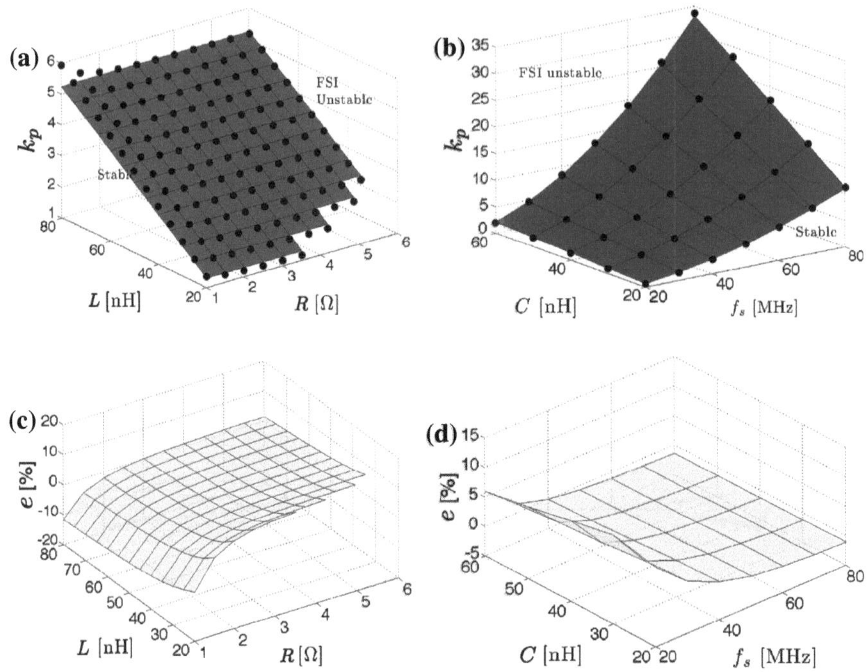

Fig. 3.7 FSI boundary surfaces obtained from the switched model (*black dots*) and from the ripple-based index condition given in Eq. 3.10 with $\rho_{crit} = 0.245$ (*mesh surface*), as a function of **a** inductance L, output resistance R and proportional gain k_p and **b** output capacitance C, the switching frequency f_s and proportional gain k_p. The error between both surfaces is shown in **c**, **d**, respectively

3.2.2 Experimental Validation

This section validates the ripple-based FSI prediction approach using an experimental prototype. While the previous section parameter values correspond to a miniaturized on-chip converter, in this section these values are chosen to facilitate the implementation of the prototype. In Appendix B it is demonstrated that by scaling both the parameters of the reactive elements and the switching frequency, the relative dynamics of the converter remains equivalent. An equivalent set of parameter values used previously and which has been used for the experimental prototype is: $V_g = 6\,\text{V}$, $R = 2.5\,\Omega$, $L = 33\,\mu\text{H}$, $C = 10\,\mu\text{F}$, $f_s = 100\,\text{kHz}$, $V_m = 1.8\,\text{V}$ and $V_{ref} = 3\,\text{V}$.

In Fig. 3.9 the waveforms of the control signal and the PWM ramp voltage just before and just after the exhibition of FSI in the experimental prototype are shown. The dynamics of the system are checked by sweeping the feedback gain k_p and the switching frequency f_s in Fig. 3.9a–d respectively. In both cases the measured value of the ripple-based index just at the boundary of stability is found to be $\rho_{crit} \approx 0.27$, which is very close to the critical value of the ripple-based index obtained from

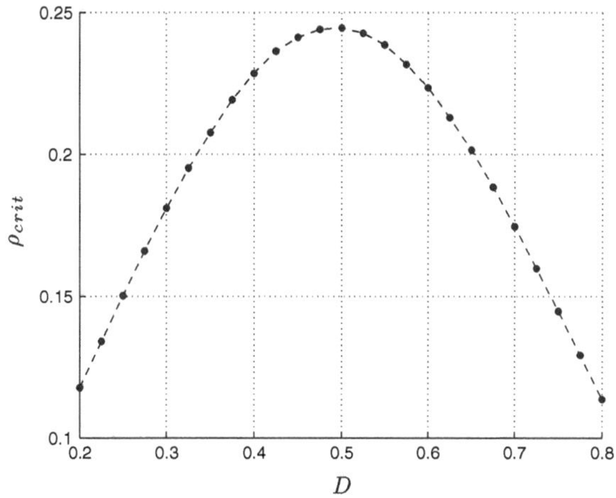

Fig. 3.8 FSI boundary curve, represented in ρ_{crit} terms and obtained from numerical simulations using the switched model, as a function of the duty cycle D

numerical simulations. Regardless of the slight difference between the experimental and the numerical value of ρ_{crit}, which can be attributed to parasitic effects, a deeper exploration of the design parameter space in Fig. 3.10 shows that the measured critical ripple ρ_{crit} for which FSI occurs remains almost the same within a wide range of the parameters.

3.3 Discrete-Time Model Stability Analysis. Relationship with the Ripple-Based Index

This section reviews the discrete-time model and its usefulness in predicting FSI boundary in DC-DC converters with the aim to validate and compare numerically the results from the ripple-based approach, but also as a starting point for giving a mathematical support for extending the hypothesis to other topologies. For this purpose, in this section it will be shown that the ripple at the modulator input can be included in the expression of the Jacobian matrix of the discrete-time model of the system.

3.3.1 The Discrete-Time Model

The stability analysis using the discrete-time approach is based on constructing a map, by sampling the dynamics of the converter at each switching period. In order to keep simple expressions, a proportional controller will be used instead of a PI controller in the analysis. In our case, as $\omega_{z1} < \omega_c$, and as mentioned in Giaouris et al. (2009), an integral term in the controller will not affect the stability boundary of

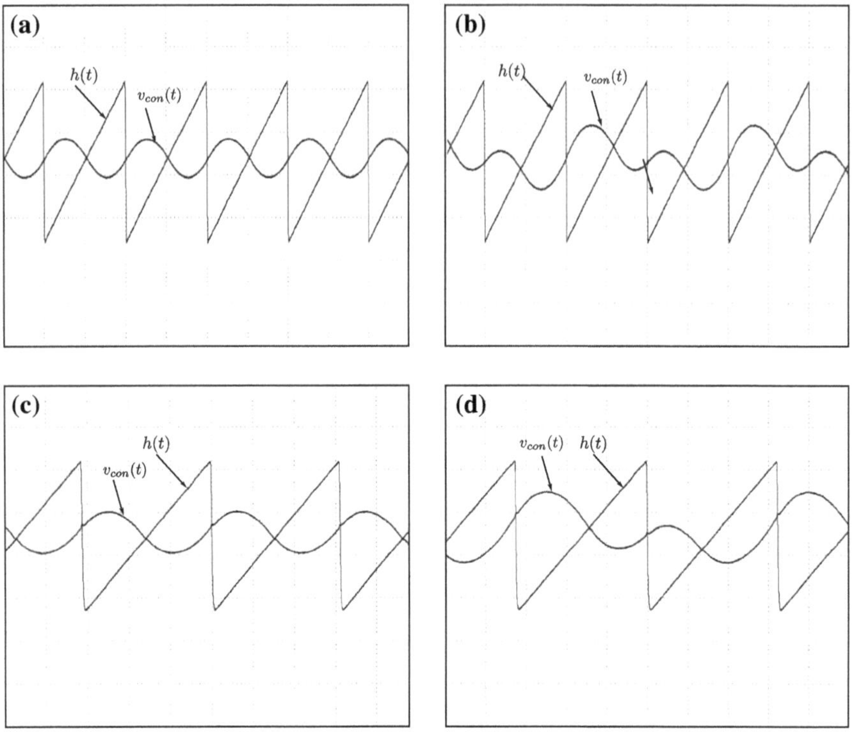

Fig. 3.9 Control signal $v_{con}(t)$ and modulator ramp $h(t)$ waveforms just before and just after FSI occurrence by sweeping **a**–**b** the proportional gain k_p and **c**–**d** the switching frequency f_s. Measured critical ripple amplitude at modulator input normalized to the ramp amplitude is $\rho_{crit} \approx 0.27$ for both cases

FSI. This is because the zero of the PI controller is placed at low frequencies, and in this case the effect of the integral term is just to correct the steady state error while the fast dynamics remains equivalent to those obtained from a proportional controller because one of the eigenvalues remains close to 1. In CCM, the dynamical behavior can be described by the following linear state equations corresponding to each state (ON and OFF) of the switch S driven by the signal $q(t)$ (See Fig. 3.5)

$$\dot{\mathbf{x}}(t) = \mathbf{A}_1\mathbf{x}(t) + \mathbf{B}_1 \quad \text{if } S \text{ is ON } (q(t) = 1)$$
$$\dot{\mathbf{x}}(t) = \mathbf{A}_2\mathbf{x}(t) + \mathbf{B}_2 \quad \text{if } S \text{ is OFF } (q(t) = 0)$$

The switch S is in the ON state at the beginning of the switching period and switches to OFF at instant d_n such that $v_{con}(d_nT) = h(d_nT)$, where d_n is the duty cycle, defined as the fraction of time during which the first configuration (ON) holds. For a linear system with constant state matrix $\mathbf{A}_k$ and input vector $\mathbf{B}_k$ and with initial time t_i and final time t_f, an exact solution can be obtained and it is given by the following expression

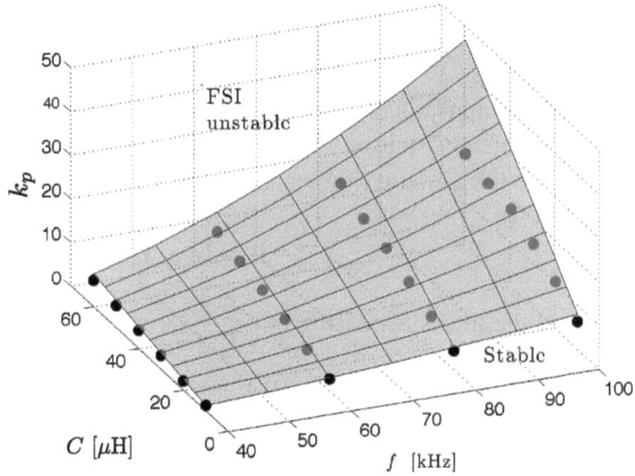

Fig. 3.10 FSI boundary surfaces obtained from experimental measurements (*black dots*) and using the ripple-based index ($\rho_{crit} = 0.245$) (*mesh surface*) over the design parameter space k_p, f_s and C

$$\mathbf{x}(t_f) = e^{\mathbf{A}_k(t_f-t_i)}\mathbf{x}(t_i) + \int_{t_i}^{t_f} e^{\mathbf{A}_k\theta}\mathbf{B}_k d\theta \tag{3.14}$$

The map $\mathbf{P}$ that relates consecutive samples, namely the state variables $\mathbf{x}_n$ at the beginning of the cycle to $\mathbf{x}_{n+1}$ at the end of the same cycle, can be obtained by stacking up the corresponding solutions during each switching cycle, and it can be expressed as El Aroudi et al. (2005):

$$\mathbf{x}_{n+1} = \mathbf{P}(\mathbf{x}_n) = \Phi\mathbf{x}_n + \Psi \tag{3.15}$$

where the matrix Φ and vector Ψ are given by

$$\Phi = \Phi_2(\bar{d}_n T)\Phi_1(d_n T),\ \Psi = \Phi_2(\bar{d}_n T)\Psi_1(d_n T) + \Psi_2(\bar{d}_n T)$$

being

$$\Phi_k(t_k) = e^{\mathbf{A}_k t_k},\quad \Psi_k(t_k) = \int_0^{t_k} e^{\mathbf{A}_k\theta} d\theta \mathbf{B}_k \text{ for } k = 1, 2 \tag{3.16}$$

and $\bar{d}_n = 1 - d_n$. Additionally, the switching condition, which depends upon the control voltage and the sawtooth modulator $h(t)$, can be expressed as

$$\sigma(d_n T) = \mathbf{K}(\mathbf{X}_{ref} - \Phi_1(d_n T)\mathbf{x}_n + \Psi_1(d_n T)) - h(d_n T) \tag{3.17}$$

where $\mathbf{K} = (k_v,\ k_i)$ is the vector of feedback gains and $\mathbf{X}_{ref} = \left(V_{ref},\ I_{ref}\right)^t$. Note that $k_v = k_p$, $k_i = 0$ and $I_{ref} = 0$ for a VMC converter with proportional compensator.

3.3.2 Stability Analysis Using the Discrete-Time Model

The stability analysis is carried out by studying the local behavior of the map in the vicinity of steady-state $\mathbf{x}^*$, thereby extracting the Jacobian matrix $\mathbf{J}$, whose eigenvalues give the amount of expansion and contraction near this fixed point El Aroudi et al. (2005):

$$\mathbf{J} = \mathbf{J}_{\mathbf{x}} - \mathbf{J}_d \mathbf{J}_{\sigma_d}^{-1} \mathbf{J}_{\sigma_{\mathbf{x}}} \tag{3.18}$$

where all the terms appearing in Eq. 3.18 are given by

$$\mathbf{J}_{\mathbf{x}} = \left.\frac{\partial \mathbf{P}}{\partial \mathbf{x}_n}\right|_{d_n=D} = \Phi_2(\overline{D}T)\Phi_1(DT) \tag{3.19}$$

which is the product of the two state transition matrices corresponding to each switching interval and which models the effect of a small change of the state variables of the system at the beginning of a switching cycle, on the state variables at the end of the same cycle.

$$\mathbf{J}_d = \left.\frac{\partial \mathbf{P}}{\partial d_n}\right|_{d_n=D,\mathbf{x}_n=\mathbf{x}^*} = \Phi_2(\overline{D}T)\Delta\dot{\mathbf{x}}T \tag{3.20}$$

which corresponds to the effect of the state transition matrix Φ_2 on the discontinuity of the vector field $\Delta\dot{\mathbf{x}}$, and which models the effect of a small change of the value of the duty cycle on the state variables at the end of the cycle.

$$\mathbf{J}_{\sigma_d} = \left.\frac{\partial \sigma}{\partial d_n}\right|_{d_n=D,\mathbf{x}_n=\mathbf{x}^*} = (-\mathbf{K}\dot{\mathbf{x}}(DT^-) - m_c)T \tag{3.21}$$

which corresponds to the difference between the slope $-\mathbf{K}\dot{\mathbf{x}}(DT^-)$ of the control signal $v_{con}(t)$ and $m_c = V_m f_s$, that of the PWM sawtooth signal $h(t)$. This is the small signal model of the original nonlinear switching condition given in Eq. 3.17. Finally,

$$\mathbf{J}_{\sigma_{\mathbf{x}}} = \left.\frac{\partial \sigma}{\partial \mathbf{x}_n}\right|_{d_n=D} = -\mathbf{K}\Phi_1(DT) \tag{3.22}$$

which models how an initial small change in the state variables affects the switching condition in Eq. 3.17 from cycle to cycle. In the previous equations, it was assumed that in steady state, $\mathbf{x}_n = \mathbf{x}^*$, $d_n = D$, $\overline{D} = 1 - D$. Besides, $\dot{\mathbf{x}}(DT^-) = \mathbf{A}_1\mathbf{x}(DT) + \mathbf{B}_1$, $\dot{\mathbf{x}}(DT^+) = \mathbf{A}_2\mathbf{x}(DT) + \mathbf{B}_2$. For the VMC buck converter, $\mathbf{x} = (v_C, \; i_L)^t$ and the matrices $\mathbf{A}_k$ and vectors $\mathbf{B}_k$ $(k = 1, 2)$ are given by

$$\mathbf{A}_1 = \mathbf{A}_2 = \begin{pmatrix} -\frac{1}{RC} & \frac{1}{C} \\ -\frac{1}{L} & 0 \end{pmatrix}, \mathbf{B}_1 = \begin{pmatrix} 0 \\ \frac{V_g}{L} \end{pmatrix}, \mathbf{B}_2 = \mathbf{0} \tag{3.23}$$

and $\Delta\dot{\mathbf{x}} = \dot{\mathbf{x}}(DT^-) - \dot{\mathbf{x}}(DT^+)$ can be obtained as follows

$$\Delta\dot{\mathbf{x}} = (\mathbf{A}_1 - \mathbf{A}_2)\mathbf{x}(DT) + (\mathbf{B}_1 - \mathbf{B}_2) \tag{3.24}$$

Since for the buck converter $\mathbf{A}_1 = \mathbf{A}_2$ and $\mathbf{B}_2 = \mathbf{0}$, the previous equation gives

$$\Delta\dot{\mathbf{x}} = \mathbf{B}_1 = \begin{pmatrix} 0 \\ \frac{V_g}{L} \end{pmatrix} \tag{3.25}$$

Moreover, for a switching buck converter under a proportional VMC, one has

$$\mathbf{K}\dot{\mathbf{x}}(DT^-) = \mathbf{K}(\mathbf{A}_1\mathbf{x}(DT) + \mathbf{B}_1) = k_p\dot{v}_C(DT^-) \tag{3.26}$$

It can be observed that the final expression of the Jacobian matrix is composed mainly by transition matrices of each configuration but it also includes information about the left and the right time derivative at the switching instant and the slope m_c of the sawtooth PWM signal $h(t)$.

3.3.3 Condition for FSI

The characteristic polynomial equation $p_{\mathbf{J}}(\lambda)$ of the Jacobian matrix $\mathbf{J}$ is given by

$$p_{\mathbf{J}}(\lambda) := \det(\mathbf{J} - \lambda\mathbf{I}) = 0 \tag{3.27}$$

In the two-dimensional system case, Eq. 3.27 becomes,

$$p_{\mathbf{J}}(\lambda) = \lambda^2 - \mathrm{tr}(\mathbf{J})\lambda + \det(\mathbf{J}) = 0 \tag{3.28}$$

where $\det(\cdot)$ and $\mathrm{tr}(\cdot)$ stand for the trace and the determinant of a matrix respectively. Eq. 3.28 can be developed as follows

$$p_{\mathbf{J}}(\lambda) := \lambda^2 - \lambda\,\mathrm{tr}(\Phi) + \lambda\mathbf{J}_{\sigma_d}\,\mathrm{tr}(\Phi_2(\overline{D}T)\Delta\dot{\mathbf{x}}T\mathbf{K}\Phi_1(DT)) + \det(\Phi) = 0 \tag{3.29}$$

where $\Phi = \Phi_2(\overline{D}T)\Phi_1(DT)$. Then, isolating the term $\mathbf{J}_{\sigma_d}$ from the previous equation, one obtains

$$\mathbf{J}_{\sigma_d} = \frac{\lambda^2 - \lambda\,\mathrm{tr}(\Phi) + \det(\Phi)}{-\lambda\,\mathrm{tr}(\Phi_2(\overline{D}T)\Delta\dot{\mathbf{x}}T\mathbf{K}\Phi_1(DT))} \tag{3.30}$$

It is known that FSI occurs when one of the eigenvalues crosses the unit circle at the point (−1,0) in the complex plane, as it was explained in Sect. 1.3.3. Therefore, the FSI boundary can be obtained by forcing $\lambda = -1$ in Eq. 3.30, hence, obtaining:

$$\mathbf{J}_{\sigma_d} = \frac{1 + \mathrm{tr}(\Phi) + \det(\Phi)}{\mathrm{tr}(\Phi_2(\overline{D}T)\Delta\dot{\mathbf{x}}T\mathbf{K}\Phi_1(DT))} \tag{3.31}$$

Using Eq. 3.21,

$$-\,(\mathbf{K}\dot{\mathbf{x}}(DT^-) + m_c)T = \frac{1 + \mathrm{tr}(\Phi) + \det(\Phi)}{\mathrm{tr}(\Phi_2(\overline{D}T)\Delta\dot{\mathbf{x}}T\mathbf{K}\Phi_1(DT))} \tag{3.32}$$

Solving Eq. 3.32 for a certain design parameter, the FSI stability boundary corresponding to this parameter can be obtained.

Having presented the nonlinear discrete-time model and its Jacobian matrix for the VMC buck converter, it is possible to compare the results obtained from it with those obtained from the ripple-based index. According to Appendix B, the whole converter design parameter space can be explored by sweeping only two parameters, namely, $\tau = f_s RC = f_s/\omega_c$ and $\Gamma = LCf_s^2 = f_s^2/\omega_0^2$. Figure 3.11 shows a comparison between the boundary surfaces obtained from both approaches in such normalized design parameter space. Note that, for low values of τ, the error increases, which is in good concordance with the results previously obtained in Fig. 3.7 stating that the ripple-based index is not accurate enough for low values of R.

3.3.4 Stability Analysis Including the Ripple Amplitude

After validating that the ripple-based index approach and the discrete-time model give similar stability boundaries in terms of FSI, this section will establish a relationship between the ripple at the input modulator and the Jacobian matrix of the discrete-time model. In Lehman and Bass (1996), the ripple of the state-variable was added to the averaged model to improve the prediction of its dynamic response. In Appendix C, it is demonstrated that the slope of the control voltage at the switching instant and the ripple amplitude are proportional and they are related by the following relationship:

$$\dot{v}_C(DT^-) = 4 f_s \hat{V}_C \tag{3.33}$$

Revisiting the different terms composing the Jacobian matrix given in Eq. 3.18, it is possible to observe that $\mathbf{J}_{\sigma d}$ contains the derivative of the state variables at the switching instant, and therefore, from Eq. 3.33, it also contains the ripple information.

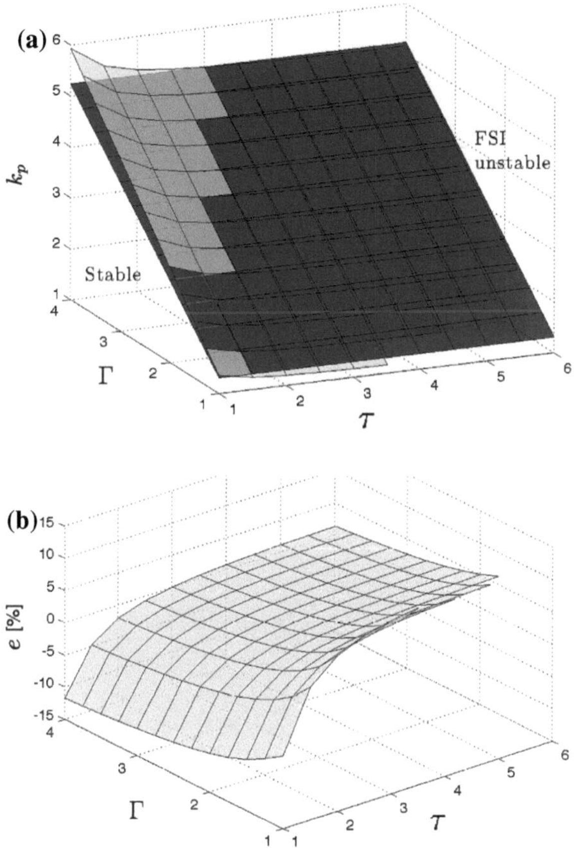

Fig. 3.11 **a** FSI boundary surfaces obtained from the discrete-time model (*white*) and from ripple-based index approach (*black*) as a function of the proportional gain k_p, Γ, and τ. **b** Error between both surfaces

Figure 3.12 shows the stability surfaces obtained from the discrete-time model and imposing FSI condition in Eq. 3.28, by using, on the one hand, the exact value of the derivative at the switching instant and on the other hand, the expression in Eq. 3.33, which includes the feedback ripple as its indirect estimate. The error between both surfaces is very small, hence demonstrating the accuracy of the ripple as an estimate of the time derivative at the switching instant. Note that the approximation error starts to increase for low values of τ but depends slightly upon Γ.

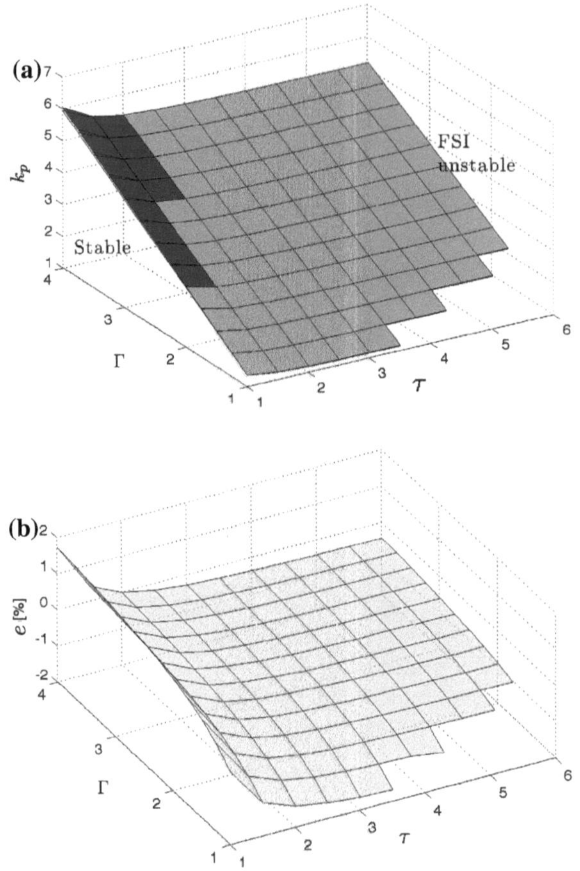

Fig. 3.12 FSI boundary surfaces obtained from the exact discrete-time model (*black*) and by approximating the derivative of the output voltage at the switching instant $\dot{v}_C(DT^-)$ for ripple amount $\hat{V}_C$ according to Eq. 3.33 (*white*) as a function of the proportional gain k_p, Γ and τ. **b** Error between both surfaces

3.4 Design-Oriented Ripple-Based Index Mathematical Demonstration

3.4.1 Revisiting the State Transition Matrices

The previous section has validated the ripple approach through contrasting the results with the discrete-time model, and also unveiled that the ripple can be included in the Jacobian matrix through its relation with the PWM switching condition. Nevertheless, it is also apparent that this approach involves sophisticated analysis, that is not of practical engineering use. On the other hand, the ripple-based approach, which has been validated in-depth for the complete design space in Sect. 3.2, is based on a starting hypothesis and lacks a solid mathematical justification. This section will derive the ripple-based FSI index approach from the discrete-time model, taking advantage of the fact that the previous section has demonstrated that the ripple of the control signal can be included in the Jacobian matrix of this model. The aim of such demonstration is to give a mathematical support to obtain a closed-form expression for the stability boundary and to examine the accuracy of such approach. The simplification of this matrix is carried out for the particular case of a VMC buck converter with a simple proportional feedback gain. Note that in the previous works (for example di Bernardo et al. (1997)), which also addressed the issue of obtaining a closed-form expression, the derived expressions are complicated since practical considerations were not used, thus making difficult their interpretation to derive useful design criteria for practical engineering use. In this section such practical considerations will be taken into account in deriving the state transition matrix, corresponding to each linear configuration used by the converter, and therefore in obtaining simplified, but accurate enough, practical expressions for the Jacobian matrix. First, the state transition matrix $\Phi_k(t)$, which relates the final state $\mathbf{x}(t_f)$ to the initial conditions $\mathbf{x}(t_i)$ during each time interval, is obtained. Using the expression of Eq. 3.14, although very accurate, does not allow to obtain clear design conditions. Instead of using an exact expression of the transition matrix, an approximated expression considering practical circuit conditions will be used. Let us first consider the solution in its general form as follows:

$$\mathbf{x}(t_f) = \mathbf{x}(t_i) + \int_{t_i}^{t_f} (\mathbf{A}_k\mathbf{x}(\theta) + \mathbf{B}_k)d\theta, \quad k = 1, 2 \tag{3.34}$$

By particularizing for the buck converter of Fig. 3.5, and considering the ON configuration ($k = 1$), the following expression of the solution can be derived:

$$\begin{pmatrix} v_C(t_f) \\ i_L(t_f) \end{pmatrix} = \begin{pmatrix} v_C(t_i) \\ i_L(t_i) \end{pmatrix} + \int_{t_i}^{t_f} \begin{pmatrix} \dfrac{i_L(\theta)}{C} - \dfrac{v_C(\theta)}{RC} \\ V_g - v_C(\theta) \end{pmatrix} d\theta \tag{3.35}$$

The simplified transition matrix will be obtained by taking into account the DC-DC buck converter circuit considerations i.e, low output voltage ripple with respect to the inductor current ripple. Note that the well-known exponential matrix simplification based on the Taylor expansion $e^{\mathbf{A}_k t} = \mathbf{I} + \mathbf{A}_k t + (\mathbf{A}_k t)^2/2! + \ldots$ does not make such a distinction between the two state variables. This circuit-based consideration allows to obtain a considerable simplification, but with high accuracy, by reducing the number of terms in the final expression of the state transition matrices involved in the Jacobian matrix. In the case of the buck converter, an approximated expression of $\Phi_k(t)$ is given by

$$\tilde{\Phi}_k(t) = \begin{pmatrix} 1 - \dfrac{t}{RC} & \dfrac{t}{C} - \dfrac{t^2}{2RC} \\ -\dfrac{t}{L} & 1 - \dfrac{t^2}{2LC} \end{pmatrix} \tag{3.36}$$

Also, $\mathbf{A}_1 = \mathbf{A}_2$, which implies $\Phi_1(t) = \Phi_2(t)$ and therefore the matrix $\Phi = \Phi_2(\overline{D}T)\Phi_1(DT)$ in Eq. 3.18 can be approximated by a simpler matrix $\tilde{\Phi}$ given by

$$\tilde{\Phi} = \begin{pmatrix} 1 - \dfrac{T}{RC} & \dfrac{T}{C} - \dfrac{T^2}{2RC} \\ -\dfrac{T}{L} & 1 - \dfrac{T^2}{2LC} \end{pmatrix} \tag{3.37}$$

Each term involved in the expression of the Jacobian given in Eq. 3.18 is reviewed according to the previous simplification. Finally, from the same expression as in Eq. 3.31, along with the expression of $\mathbf{J}_{\sigma_d}$ given in Eq. 3.21, a closed-form expression for predicting FSI boundary is obtained:

$$\mathbf{J}_{\sigma_d} = -k_p \dot{v}_C(DT^-)T - m_c T = -\frac{k_p V_g \mathcal{P}_v}{4LCf_s^2} \tag{3.38}$$

where $\mathcal{P}_v$ is given by the following expression

$$\mathcal{P}_v = \frac{2\omega_0^2\omega_c T^3 D^2 \bar{D}^2 + 4T^2(\omega_c^2 - \omega_0^2)D\bar{D}^2 + 8 - 4T\omega_c}{T^4\omega_0^4 D^2\bar{D}^2 - 4T^2(\omega_c^2 - \omega_0^2)D\overline{D} + 8 - 4T\omega_c}$$

Furthermore, the expression given in Eq. 3.38 can be rewritten as a function of the output voltage ripple $\hat{V}_C$ taking into account the relationship between the derivative at the switching instant and such ripple magnitude given in Eq. 3.33:

$$-4k_p\hat{V}_C - V_m = -\frac{k_p V_g \mathcal{P}_v}{4LCf_s^2} \tag{3.39}$$

3.4.2 Critical Ripple Expression

The closed-form expression given in Eq. 3.39 still lacks a design-oriented standpoint. With the aim of obtaining a simpler expression, the ripple amplitude is approximated as in Eq. 3.11, which has been validated for the complete design parameter space in Appendix D, along with considering $\mathcal{P}_v = 1$. Then, Eq. 3.39 can be rewritten as:

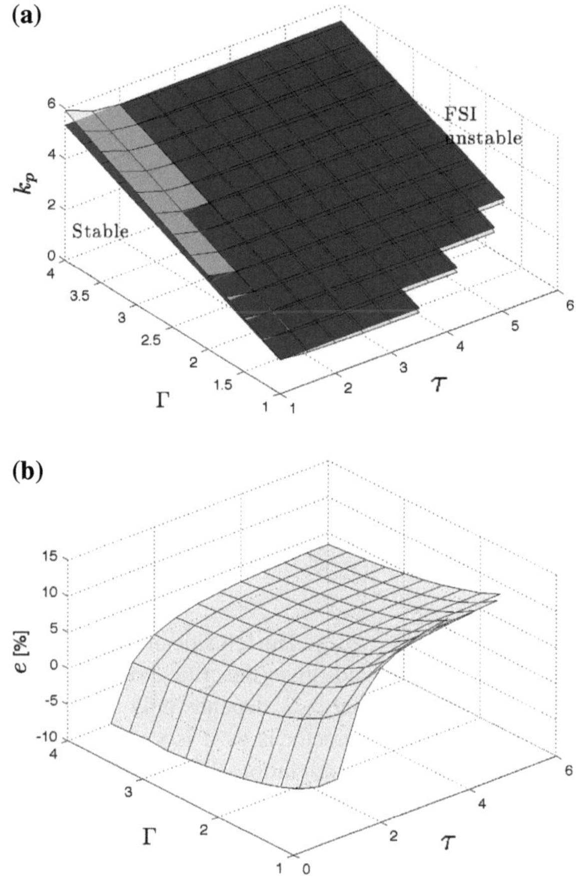

Fig. 3.13 **a** FSI boundary surfaces obtained from the discrete-time model (*white*) and from the ripple-based condition given in Eq. 3.42 $\mathcal{P}_v = 1$ (*black*) as a function of the proportional gain k_p, Γ and τ. **b** Error between both surfaces

$$\frac{k_p V_g}{4LCf_s^2} - 4k_p \hat{V}_C = V_m \tag{3.40}$$

$$\frac{k_p V_g}{4V_m LCf_s^2}(1 - 2D\overline{D}) = 1 \tag{3.41}$$

To make this expression consistent with the ripple-based approach given in Eq. 3.12, let us multiply both sides of Eq. 3.41 by $D\overline{D}$ and divide them by $2(1 - 2D\overline{D})$, thereby obtaining the following final stability condition for predicting FSI occurrence:

$$\underbrace{\frac{k_p V_g D\overline{D}}{8V_m LCf_s^2}}_{\rho} = \underbrace{\frac{D\overline{D}}{2 - 4D\overline{D}}}_{\rho_{crit}} \tag{3.42}$$

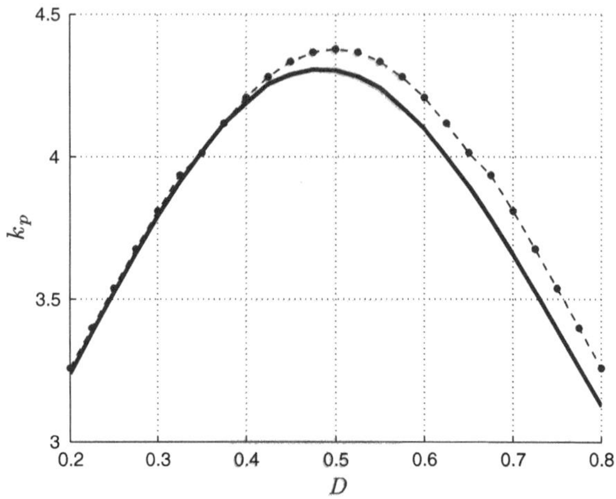

Fig. 3.14 FSI boundary curves obtained from the discrete-time model (Eq. 3.18, *dots*) and from ripple-based condition given in Eq. 3.42 (*solid*) with $\mathcal{P}_v = 1$ as a function of the duty cycle D and the proportional gain k_p with $\tau = 2.5$ and $\Gamma = 3.3$

The system will be stable if $\rho < \rho_{crit}$. The left side of this equation is the defined ripple-based index ρ while its right side gives the critical ripple value ρ_{crit} for which the system will exhibit FSI. Note that ρ_{crit} depends only on the duty cycle D and its dependence in terms of this parameter is as it is shown previously in Fig. 3.8 which was obtained from numerical simulations. The stability curve obtained from the exact Jacobian matrix given in Eq. 3.18 together with those obtained from the ripple-based index condition given in Eq. 3.42 are shown in Fig. 3.13 by sweeping the parameters k_p, τ and Γ. The results show that the error produced by the approximation is negligible in a wide range of the design parameter space. The stability curve obtained from the exact Jacobian matrix and the approximated Jacobian with $\mathcal{P}_v = 1$ by sweeping D and calculating the critical value of the feedback gain k_p is depicted in Fig. 3.14 showing the accuracy of the closed-form expression for the whole practical range of duty cycles.

3.4.3 Ripple-Based Index Approach Limitations

The previous section has given a consistent mathematical demonstration of the ripple-based index approach for predicting FSI. However, this approach is not accurate enough in predicting the FSI boundary when some design parameters, namely Γ and τ, are relatively low, as it is shown in Fig. 3.13. This is considered to be a penalty to having a simplified expression with the important advantage of being oriented to design. This total error due to the different approximations done in the process of obtaining the closed-form expression for predicting FSI stands from different error sources. First, the ripple has been included in the discrete-time model as an estimate of the derivative of the state variable at the switching instant. As it can be observed in Appendix C, τ is the parameter that produces a major error (although it is only of 1 %). Note that τ is related to the output load of the converter, so that low values of τ ($\tau << 1$) indicates that it is no more valid the assumption, which has been used for relating the output voltage ripple to the derivative of such signal at the switching instant, that the inductor ripple current flows mostly through the output capacitor, hence notably increasing the error when using the ripple-based approach. On the other hand, in order to keep the design-oriented formulation, the ripple has been approximated by Eq. 3.11. In Appendix D (Fig. D.2) it has been demonstrated that both parameters have an impact on this approximation, but the error is very low (< 2 %). Finally, the last approximation has been done by simplifying the Jacobian matrix using circuit considerations. This is the approximation which introduces more error, since both Γ and τ have an important impact on the approximation of the discrete-time model and when these parameters are low ($\Gamma < 1$ and $\tau < 1$), the error increases as it is shown in Fig. 3.13 and starting to lose the approach validity.

3.5 A Design-Oriented Combined Approach for Predicting Overall Stability Boundaries

The previous section has demonstrated that the voltage ripple amount at the PWM modulator input allows to quantitatively predict the FSI exhibition in a buck DC-DC converter. This approach compresses the complete multidimensional parameter space into a single index, thereby providing a design-oriented parameter space characterization in terms of FSI. It has been shown that, independently of the parameter varied, when the ripple at the PWM modulator exceeds a critical value, the system exhibits s. In this section, the approach is combined with the SSI index given in Eq. 3.8, in order to completely divide and classify the design parameter space in different stability areas. The combined design-oriented stability conditions are:

- Stability condition for avoiding SSI occurrence:

$$\rho_{SS} := \frac{k_p}{V_m} V_g \left(\frac{\omega_{z1}}{\omega_{RC}} - 1 \right) < 1 \tag{3.43}$$

- Stability condition for avoiding FSI occurrence:

$$\rho_{FS} := \frac{k_p}{V_m} \frac{V_g D(1-D)}{8LCf_s^2} < \rho_{crit}(D) \tag{3.44}$$

The design-oriented stability map within the parametric space composed of ω_{z1}, k_p, f_s and R, V_m, L is shown in Fig. 3.15. The results obtained from the concurrent application of both instability indexes aiming to predict both SSI and FSI are super-

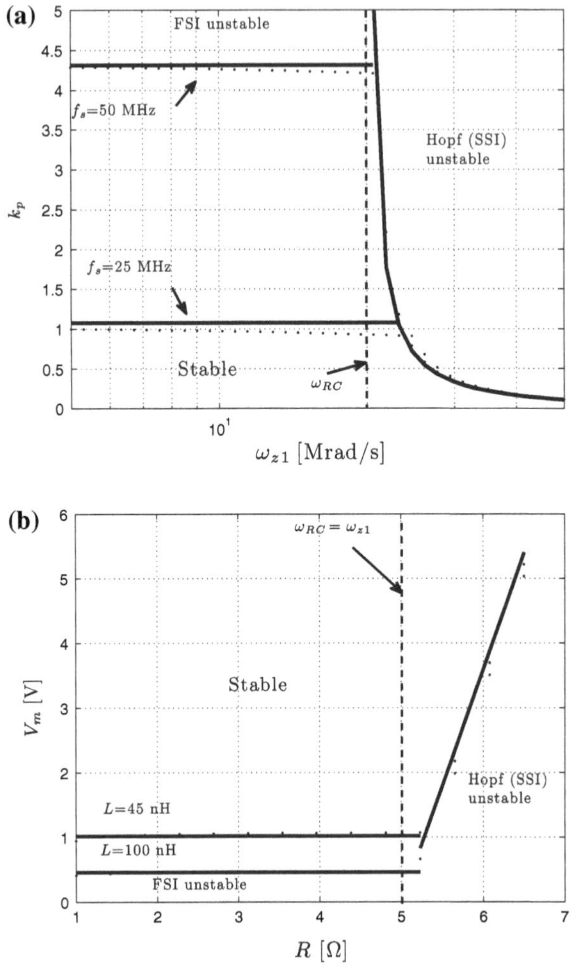

Fig. 3.15 Overall stability map obtained from the discrete-time model (*dash-dots*) compared with design-oriented models: SSI boundary (*solid blue*) from Eq. (3.43) and FSI boundary (*solid green*) from ripple-based index given in Eq. (3.44) with $\rho_{crit}(0.5) = 0.245$. The obtained results as a function of **a** ω_{z1}, k_p and f_s **b** R, V_m and L

imposed and show a strong correlation to those obtained from the conventional and mature method based on the discrete-time model, hence validating the approach.

3.6 Extension to Discontinuous Conduction Mode

The characterization of dynamics and the effect of the parameters on the stability boundary have been addressed in Sect. 2.1 showing that, in VMC buck converter working in DCM and under a PI controller, the dependence of the stability boundary upon the system parameters is more complex. The derivation of a closed-form stability condition for the VMC buck converter working in DCM has been previously tackled in (Tse, 1994a) but the work assumes, a sample and hold action in the feedback to simplify the analysis. Therefore alternatively to this work, this section attempts to extend the stability analysis from a ripple-based perspective to the DCM case, hence illustrating the general purpose applicability of this index and obtaining a common framework to compare both conduction modes in terms of stability. Therefore, according to the ripple hypothesis given in Eq. 3.10 the stability condition for a VMC buck converter working in DCM is:

$$\rho_{\mathrm{DCM}} := \frac{k_p}{V_m} \hat{V}_{C,\mathrm{DCM}} < \rho_{crit,\mathrm{DCM}} \tag{3.45}$$

The condition for the system to be working in this mode can be expressed as a function of the converter parameters as:

$$K_{\mathrm{DCM}} < 1 - M \tag{3.46}$$

where $M = V_o/V_g$ is the voltage conversion ratio and K_{DCM} is given by:

$$K_{\mathrm{DCM}} := \frac{2Lf_s}{R} \tag{3.47}$$

The stability analysis in terms of the ripple-based index is more complex than in the CCM case, since the ripple in DCM has a more complex expression than in CCM (Erickson and Maksimovič, 2001):

$$\hat{V}_{C,\mathrm{DCM}} := \frac{\left(\frac{V_g - V_o M}{L f_s} - \frac{V_o}{R}\right)^2}{2C(V_g - V_o)ML} \tag{3.48}$$

Moreover, the duty cycle also depends upon K_{DCM}:

$$D_{\mathrm{DCM}}(K_{\mathrm{DCM}}, M) = M\sqrt{\frac{K_{\mathrm{DCM}}}{1-M}} \tag{3.49}$$

Note that at $K_{\mathrm{DCM}} = 1 - M$ (boundary between DCM and MCC) the duty cycle coincides with the conversion ratio M.

The following simulation addresses the validation of the ripple-index approach for DCM and the disclosure of the ρ_{crit} dependence. The parameters in this section are, as in the previous sections, oriented to miniaturization: $V_g = 6\,\mathrm{V}$, $V_{ref} = 3\,\mathrm{V}$, $R = 6\,\Omega$, $L = 15\,\mathrm{nH}$, $C = 50\,\mathrm{nF}$, $f_s = 50\,\mathrm{MHz}$, $V_m = 1\,\mathrm{V}$, $k_p = 2.5$ and $\omega_{z1} = 1\,\mathrm{Mrad/s}$.

Note that the parameters that appear in the ripple expression but not in K_{DCM} (Eq. 3.47) neither in D_{DCM} (Eq. 3.49) are the output capacitance C, the proportional

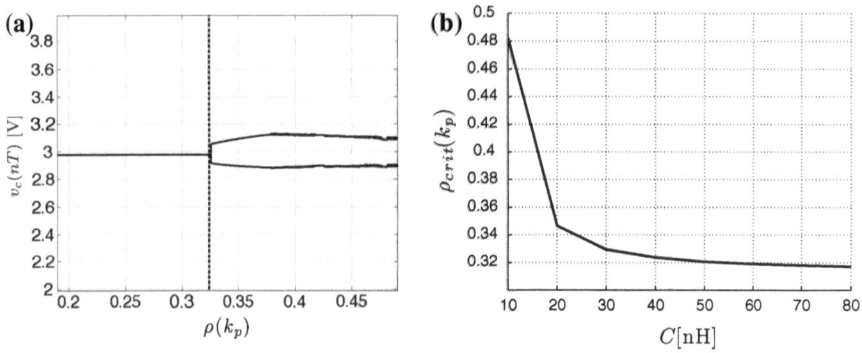

Fig. 3.16 **a** Bifurcation diagram by sweeping the proportional gain k_p and **b** the FSI boundary $\rho_{crit}(k_p)$ as a function of output capacitance C

gain k_p and the modulator ramp amplitude V_m. The characterization in Fig. 3.16 shows that, by sweeping k_p or C, the stability boundary, expressed in ripple terms ρ, remains almost constant for high C values while the stability boundary $\rho_{crit,\text{DCM}}$ improves for low values of C.

Furthermore, Fig. 3.17 demonstrates that by keeping K_{DCM} constant, despite using different values of L, f_s or R, the stability boundary is kept constant, expressed in ripple terms, and comparing with results in Fig. 3.16. A similar characterization case has been carried out by keeping constant the voltage conversion ratio, but varying the input and output voltages. Further extension is carried out by using the results from the discrete-time model derived in Appendix A showing that the stability boundary remains almost invariable, hence demonstrating that by keeping the duty cycle $D_{\text{DCM}}(K_{\text{DCM}}, M)$ constant, the stability boundary $\rho_{crit,\text{DCM}}$ is constant, which agrees with the results in CCM.

Finally, once it has been demonstrated that the stability boundary $\rho_{crit,\text{DCM}}$ is not modified provided that K_{DCM} and M are kept constant, the simulation in Fig. 3.18 shows the effect of both variables upon such boundary by means of the discrete-time model.

In Fig. 3.18, it can be observed that the effect of K_{DCM} is different from that of M. On the one hand, K_{DCM} tends to increase the stability region by increasing the critical ripple $\rho_{crit,\text{DCM}}$, whereas the voltage conversion ratio M tends to decrease the stability region by reducing such critical boundary. Besides that, there is a discontinuity between CCM and DCM, which has been already pointed out in Sect. 2.1.

The validation of the ripple index approach has been carried out in this section for the DCM, but its benefits are limited by the fact that the duty cycle depends upon many parameters of the converter, as it is shown in Eq. 3.49, and that there is a discontinuity in the stability boundary between both conduction modes, hence entailing that the stability analysis is more complex. Finally, the ripple-based index stability condition for avoiding FSI exhibition working in DCM, as a function of system parameters, can be expressed as:

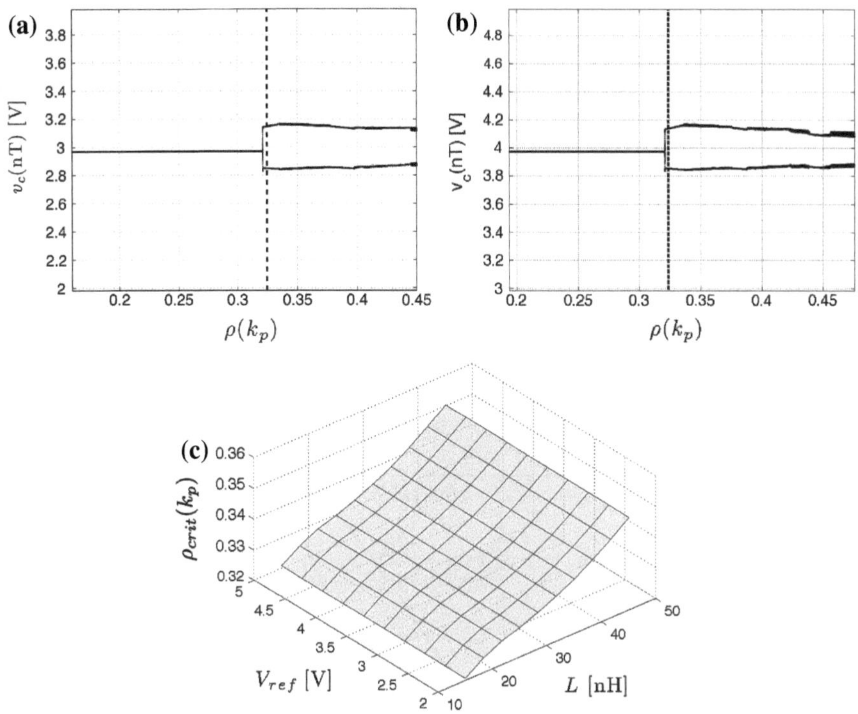

Fig. 3.17 Bifurcation diagrams by sweeping the proportional gain k_p with **a** $K_{\mathrm{DCM}} = 0.375$, as in Fig. 3.16, but with different parameter values $L = 45\,\mathrm{nH}$ $f_s = 25\,\mathrm{MHz}$ $R = 6\,\Omega$ **b** the same $M = 0.5$, but with different parameters values $V_g = 8\,\mathrm{V}$ $V_{ref} = 4\,\mathrm{V}$. **c** FSI boundary obtained from the discrete-time model by concurrently sweeping the inductance value L and the switching frequency $f_s = K_{\mathrm{DCM}}R/(2L)$, hence keeping constant K_{DCM}, and concurrently sweeping the voltage reference value V_{ref} and the input voltage $V_g = V_{ref}/D$, hence keeping constant M

$$\rho_{\mathrm{DCM}}(L, C, R, V_g, V_{ref}, V_m, f_s, k_p) < \rho_{crit,\mathrm{DCM}}(D_{\mathrm{DCM}}(K_{\mathrm{DCM}}, M)) \qquad (3.50)$$

3.7 Extension to Full-State-Feedback Controller

Modeling of CMC switching power converters has been widely explored in the past (Ridley, 1989), (Deane, 1992), (Bryant and Kazimierczuk, 2005) due to their propensity to exhibit FSI (Redl and Novak 1981). Most of the derived models take advantage of the fact that the regulator can be modeled as a first-order system, hence allowing to derive, considering piecewise linear approximations of the inductor current, a simple closed-form expression to predict the stability boundary.

In an elementary CMC converter, considering only the current loop, a well-known condition for predicting the exhibition of FSI can be written in a simple equation as a function of the slope before and after the switching instant, m_1 and m_2 respectively,

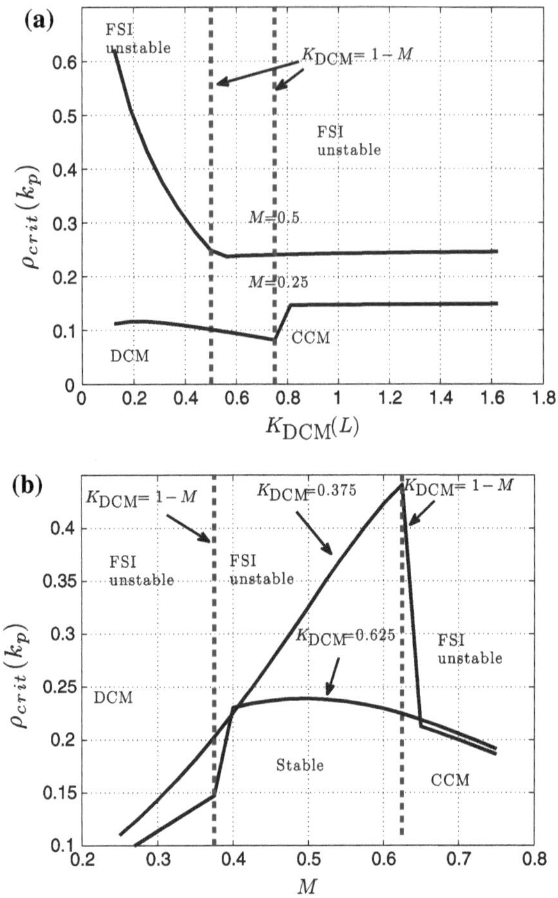

Fig. 3.18 Stability curve $\rho_{crit,\mathrm{DCM}}(k_p)$ as a function of **a** K_{DCM}, for $M = 0.5$ and $M = 0.25$ and **b** voltage conversion ratio M, for $K_{\mathrm{DCM}} = 0.625$ ($L = 25\,\mathrm{nH}$, $f_s = 50\,\mathrm{MHz}$, $R = 4\,\Omega$) and $K_{\mathrm{DCM}} = 0.375$ ($L = 15\,\mathrm{nH}$, $f_s = 50\,\mathrm{MHz}$, $R = 4\,\Omega$)

and the external added ramp slope m_c (Erickson and Maksimovič, 2001):

$$\frac{k_i m_2 - m_c}{k_i m_1 + m_c} < 1 \tag{3.51}$$

For the case of a buck converter, assuming triangular inductor current waveform, the slopes can be approximated as:

$$m_1 \simeq \frac{V_g \overline{D}}{L} \tag{3.52}$$

$$m_2 \simeq \frac{V_g D}{L} \tag{3.53}$$

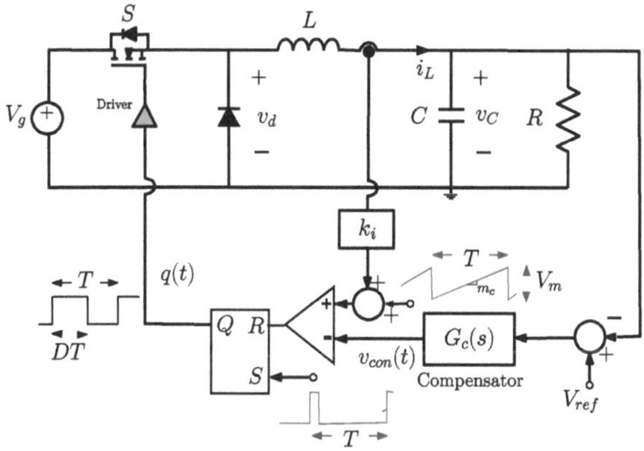

Fig. 3.19 CMC buck converter with voltage feedback loop

Then, the stability condition can be expressed as:

$$\frac{k_i m_2 - m_c}{k_i m_1 + m_c} < 1 \Rightarrow k_i \frac{V_g(2D-1)}{2L} < m_c \tag{3.54}$$

Note that without external ramp ($m_c = 0$), Eq. (3.54) implies that $m_1 > m_2$ and that the maximum duty cycle for ensuring stability is $D = 0.5$. With an external ramp and in the worst case, namely $D = 1$, the minimum required ramp slope to avoid FSI exhibition is $m_c = k_i V_g/2L$.

Regardless the simplicity of the obtained stability conditions in both cases, most of the works neglect in the system dynamics that CMC requires a voltage feedback loop in order to properly regulate the output voltage under load changes, which indeed can produce that FSI occurs even for duty cycles below 0.5, as it was pointed out in (Ridley, 1989). The main aim of this section is to explore the combination of both current/voltage loops and give a closed-form expression to predict the FSI boundary, without abandoning the ripple perspective proposed in this book. A CMC buck converter with a voltage feedback loop is shown in Fig. 3.19.

Following a similar procedure to that of the case of a VMC buck converter in Sect. 3.4, but considering in this case, the additional current feedback path in Eq. 3.21 and Eq. 3.22, namely including the proportional current gain in the vector gain $\mathbf{K} = [k_v \ k_i]$, it is possible to derive a closed-form expression for predicting FSI boundary in a similar way as in Eq. 3.38, composed in the left side of the derivative just before the switching instant of the feedback variables:

$$J_{\sigma_d} = -k_v \dot{v}_C(DT^-) - k_i \dot{i}_L(DT^-) = \frac{-k_v V_g \mathcal{P}_v}{4LCf_s} - \frac{k_i V_g \mathcal{P}_i}{2L} + m_c \tag{3.55}$$

where $\mathcal{P}_v$ is the same as in Eq. 3.39 and $\mathcal{P}_i$ is:

$$\mathcal{P}_i = \frac{2T^2\omega_0^2 D^2\overline{D}^2 + 2T^3\omega_{RC}\omega_0^2 D^2\overline{D} + 4T^2(\omega_0^2 - \omega_{RC}^2)D\overline{D} + 8 - 4T\omega_{RC}}{T^2\omega_0^2 D^2\overline{D}^2 + 4T^2(\omega_0^2 - \omega_{RC}^2)D\overline{D} + 8 - 4T\omega_{RC}} \tag{3.56}$$

Note that J_{σ_d} is composed of two kinds of terms, one depending on the inductor current and the other one on the capacitor voltage, hence increasing the complexity of the critical stability boundary. Therefore, with the aim of simplifying the final expression, Eq. (3.55) can be rewritten as:

$$k_v \dot{v}_C(DT^-) + k_i \dot{i}_L(DT^-) + m_c \simeq \frac{k_v V_g}{4LCf_s} + \frac{k_i V_g}{2L} \tag{3.57}$$

where $\mathcal{P}_v$ and $\mathcal{P}_i$ have been approximated by 1. Then, the voltage and current terms are respectively joined in both sides of the equation and the derivative of the output voltage at the switching instant is replaced by the output voltage ripple $\hat{V}_C$:

$$-k_v 4 f_s \hat{V}_C + \frac{k_v V_g}{4LCf_s} = m_c + k_i m_1 - \frac{k_i V_g}{2L} \tag{3.58}$$

which by dividing by the switching frequency f_s and rearranging terms gives

$$\frac{k_v V_g}{4LCf_s^2}\left(1 - 2D\overline{D}\right) = \frac{m_c}{f_s} + \frac{k_i}{f_s}\left(m_1 - \frac{V_g}{2L}\right) \tag{3.59}$$

Finally, it is possible to derive a unified stability condition for predicting the FSI stability boundary valid for the combined voltage and current loops. The stability condition can be expressed as:

$$\rho_v < \rho_{crit,\text{CMC}} \tag{3.60}$$

where

$$\rho_{crit,\text{CMC}} = \frac{\rho_{crit,\text{VMC}}}{V_m}\left(V_m + \frac{k_i}{f_s}\left(m_1 - \frac{V_g}{2L}\right)\right) \tag{3.61}$$

being ρ_v and $\rho_{crit,\text{VMC}}$ the ripple-based index and its critical value, respectively for the VMC case, rewritten here:

$$\rho_v = \frac{k_v}{V_m}\frac{V_g D\overline{D}}{8LCf_s^2} \tag{3.62}$$

$$\rho_{crit,\text{VMC}} = \frac{D\overline{D}}{2 - 4D\overline{D}} \tag{3.63}$$

The expression can be particularized for a VMC buck converter, by forcing $k_v = k_p$, $k_i = 0$ and considering the PWM ramp $V_m \neq 0$, or for a CMC buck converter by adjusting current gain, feedback voltage gain, and the external ramp

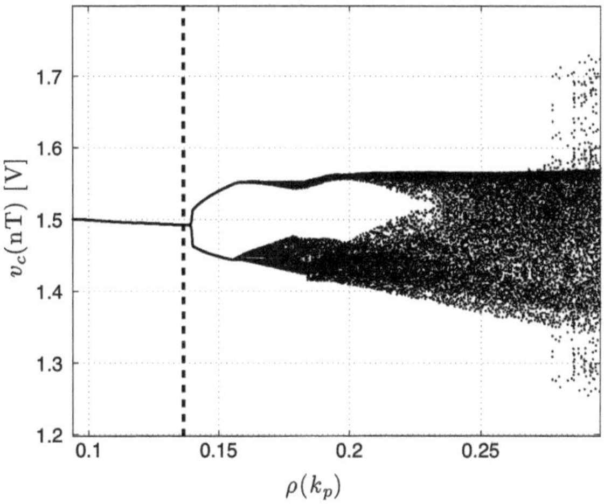

Fig. 3.20 Bifurcation diagram by sweeping k_p in a CMC buck converter with $k_i = 2$, $m_c = 0$ and $V_{ref} = 1.5$ V. $\rho_{crit,\mathrm{CMC}} = 0.1364$ obtained from Eq. 3.61

($V_m = 0$ without external ramp). The validation of the accuracy of the new design-oriented stability condition in CMC, with voltage loop but without external ramp, is shown in Fig. 3.20. This figure shows that FSI can be exhibited even for duty cycles lower than 0.5, depending on the amount of voltage ripple in the modulator, which is not possible to be predicted by the stability condition of Eq. 3.51. Furthermore, in Fig. 3.21 the stability condition is also validated with an external ramp by sweeping the current gain k_i. The parameters used in this section are the same as for the pure VMC case: $V_g = 6$ V, $V_{ref} = 1.5$ V, $R = 2.5\,\Omega$, $L = 66$ nH, $C = 20$ nF, $f_s = 50$ MHz and $V_m = 1$ V.

The stability condition implies that the current loop modifies the critical stability boundary and that by increasing the effect of the current slope m_1, for instance by increasing the current gain k_i, the FSI stability boundary is shifted away. However, this improvement upon the stability boundary will strongly depend upon the duty cycle D, as it is shown in Fig. 3.22 and Fig. 3.23. For low values of the duty cycle, there is a stability improvement compared to the VMC case, although it is reduced when the duty cycle D increases. The higher the duty cycle is, the lower the output voltage ripple is required to keep the stable behavior and for $D > 0.5$ the critical boundary $\rho_{crit,\mathrm{CMC}}$ is negative, hence being imperative to consider an external ramp to obtain a stable period-one behavior.

The obtained expression from Eq. 3.60 and Eq. 3.61 when no voltage feedback loop is considered, hence k_v=0 and m_1 given in Eq. 3.52, is:

$$\frac{k_i}{m_c}\frac{V_g}{2L} - m_1 < 1 \Rightarrow k_i\frac{V_g(2D-1)}{2L} < m_c \tag{3.64}$$

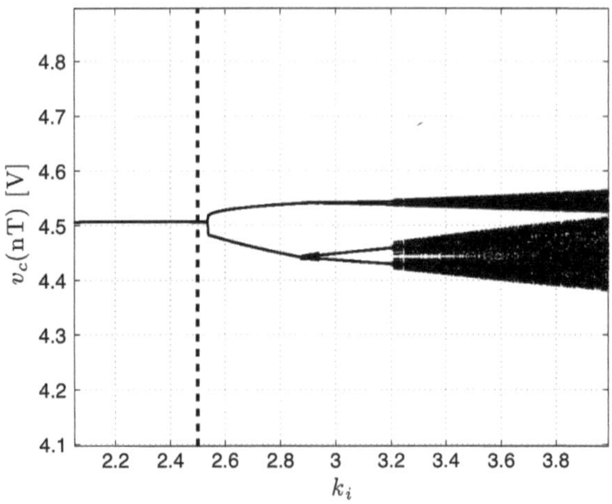

Fig. 3.21 Bifurcation diagram by sweeping k_i with $V_{ref} = 4.5$ V, $k_p = 3$ and $m_c = 100$ V/s with $V_m = 2$ V and $f_s = 50$ MHz. From the CMC stability condition in Eq. 3.60 bifurcation occurs at $k_i = 2.5$

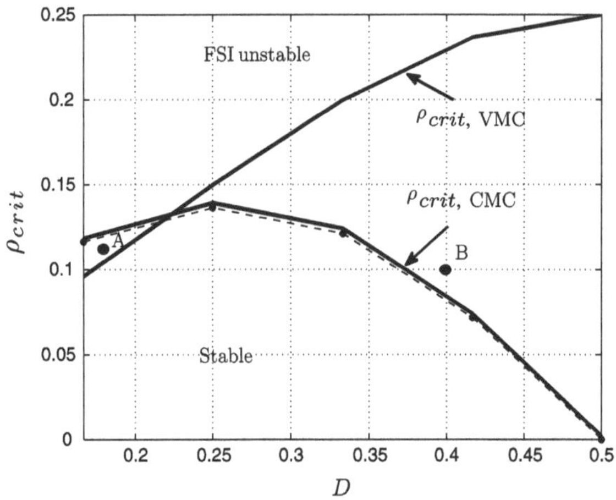

Fig. 3.22 Stability boundary by sweeping the duty cycle D in a CMC. Results obtained from the discrete-time model (*dots*) and from the design-oriented approach of Eq. (3.60) (*solid*) with $k_i = 2$. Additionally, the stability boundary in VMC $\rho_{crit,\text{VMC}}$ obtained for $k_i = 0$ is shown. A and B points are simulated and shown in Fig. 3.23

Despite the mathematical formulation is different, the results are the same as the ones obtained in Eq. 3.54. Note that the term $V_g/(2L)$ can be interpreted as a particular case of m_2 for $D = 0.5$, which coincides with the maximum value that system will be stable according to the simple stability condition given in Eq. (3.51).

Furthermore, apart from the general applicability of the stability condition given in Eq. (3.60) to VMC or CMC and with and without external ramp, the new stability

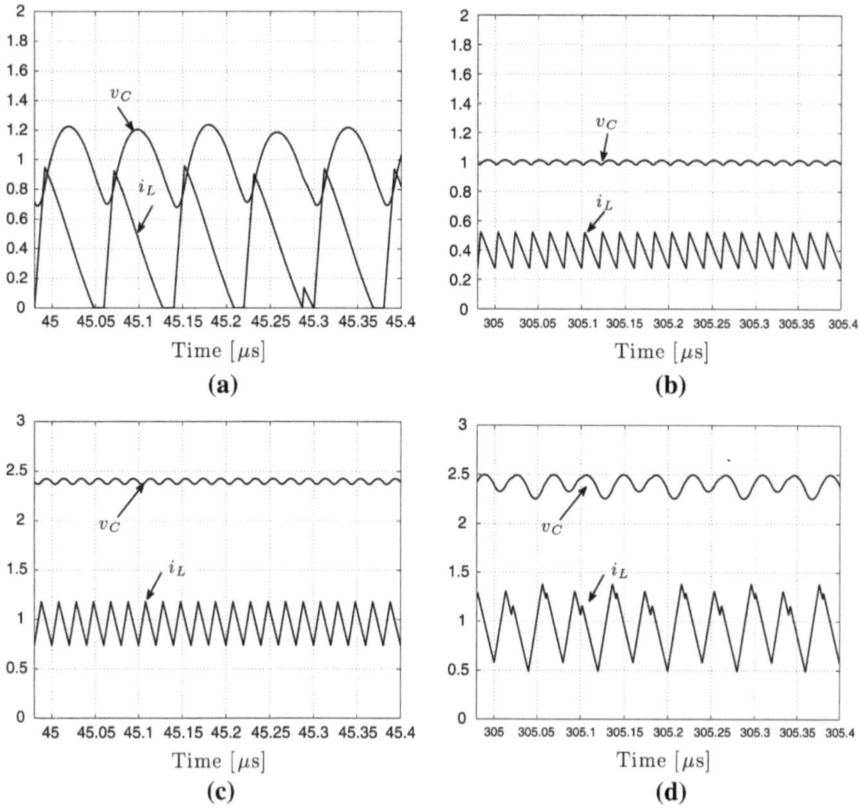

Fig. 3.23 Time simulation in VMC buck converter ($k_i = 0$ and $m_c = V_m f_s$) **a**, **c** and in CMC buck converter ($k_i = 2$ and $m_c = 0$) **b**, **d** respectively. Conditions are those indicated in point A and point B in Fig. 3.22

condition also allows to unveil the effect of converter parameters in all these different cases. For instance, parameters such as the proportional gain k_p or the output capacitor C, which only modify the voltage ripple, will clearly lead to FSI exhibition when the variation of these parameters leads to increase the voltage ripple in the feedback loop. Moreover, under no external ramp condition ($V_m = 0$), the inductance will not have any effect upon FSI stability boundary since it appears in both sides of the stability condition with the same proportion (this fact also occurs in the case of input voltage V_g) while the switching frequency will have more impact on the ripple than on the current slope, and hence decreasing its value may lead to the occurrence of FSI.

This section has extended the ripple-based index for predicting FSI boundary to the CMC buck converter, by giving a simple closed-form stability condition hence demonstrating that this condition has a dependency on the amount of the voltage ripple added by the voltage-feedback loop apart from the inductor slopes.

3.7.1 Extension to the PID Compensator

The PID compensator adds a phase lead, hence improving stability margins regarding SSI. The PID compensator description in the time and frequency domains are:

$$v_{con}(t) = k_p v_e(t) + k_d \dot{v}_e(t) + k_I \int_{-\infty}^{t} v_e(\theta) d\theta \tag{3.65}$$

$$G_c(s) = \frac{v_{con}(s)}{v_e(s)} = G\frac{(s+\omega_{z1})(s+\omega_{z2})}{s} \tag{3.66}$$

where $v_e(t) = V_{ref} - v_C(t)$ is the output voltage error. The equivalence between time domain coefficients and frequency domain coefficients is:

$$k_p = G(\omega_{z1} + \omega_{z2}) \tag{3.67}$$

$$k_d = G \tag{3.68}$$

$$k_I = G\omega_{z1}\omega_{z2} \tag{3.69}$$

Starting from the time domain description, it is possible to rewrite the control voltage as a function of the capacitor voltage and the inductor current, since $\dot{v}_e(t)$ indeed includes the inductor current, hence allowing a similar analysis to that carried out in the precedent section:

$$\dot{v}_e = -\dot{v}_C = -\frac{i_L}{C} + \frac{v_C}{RC} \tag{3.70}$$

Thus, the feedback gains k_v and k_i which multiply the feedback variables v_C and i_L respectively are:

$$k_v = k_p - \frac{k_d}{RC} = G(\omega_{z1} + \omega_{z2} - \omega_{RC}) \tag{3.71}$$

$$k_i = \frac{k_d}{C} = \frac{G}{C} \tag{3.72}$$

Note that, from the previous analysis in Eq. 3.71, it is possible to derive that the second zero ω_{z2} should remain above ω_{RC} in order to obtain $k_v > 0$, thus ensuring the negative feedback regulation, but below $\omega_0 = 1/\sqrt{LC}$ in order to obtain the desired improvement in the phase margin.

The addition of a derivative-related second zero in the compensator, apart from improving the phase margin and therefore the stability in terms of SSI, can also improve the FSI stability boundary by analogously adding the equivalent to a current term to the feedback, regardless the fact that it will have a strong dependency upon the duty cycle of the converter as it is shown in Fig. 3.24. This figure also demonstrates the validity of the previous design-oriented equation for full-state-feedback, given

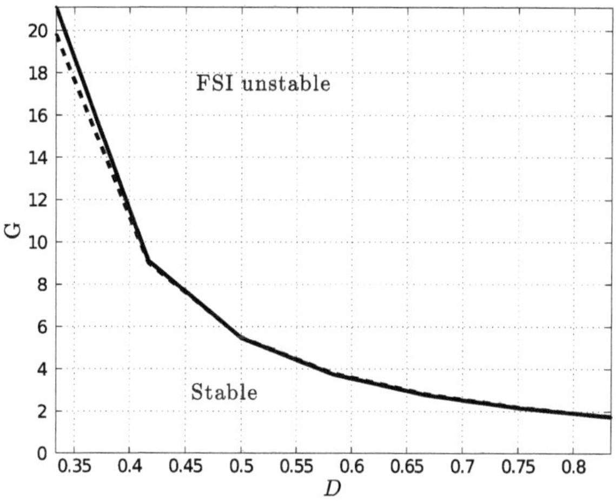

Fig. 3.24 Stability curve as a function of G by sweeping the duty cycle D. The results are obtained from the discrete-time model (*dashed*) and from design-oriented equation (*solid*) given in Eq. 3.60

in Eq. 3.60, for the case of a PID compensator including the external ramp of the PWM modulator.

From a design standpoint, it is also interesting to know the effect of the second pole that appears in a realistic PID, the s-domain representation of which can be expressed as:

$$G_c(s) = G\frac{(s+\omega_{z1})(s+\omega_{z2})}{s(s+\omega_{p2})} \tag{3.73}$$

The second pole ω_{p2} is usually limited in order to reduce the effect of noise in the circuit. Note that for $\omega_{z1} < \omega < \omega_{z2}$ and $\omega > \omega_{p2}$, G is equivalent to a proportional gain.

In Fig. 3.25, the effect of ω_{p2} on the FSI stability boundary is shown as a function of the switching frequency f_s and the duty cycle D. The stability curve has been obtained by sweeping the switching frequency or the duty cycle and obtaining the value of G_{crit} that leads to FSI exhibition. The FSI boundary is represented by $\rho_{crit,\text{PID}}$, which is defined as:

$$\rho_{crit,\text{PID}} = \frac{G_{crit}}{V_m}\frac{V_g D\overline{D}}{8LCf_s^2} \tag{3.74}$$

From these results, it can be observed that when the switching frequency is fixed well above the second pole ω_{p2}, the stability boundary in a proportional VMC buck converter is asymptotically obtained. If the switching frequency is fixed well below such pole, then the stability boundary can be derived from the previous PID case (without additional pole), thereby strongly depending on the duty cycle and the current gain.

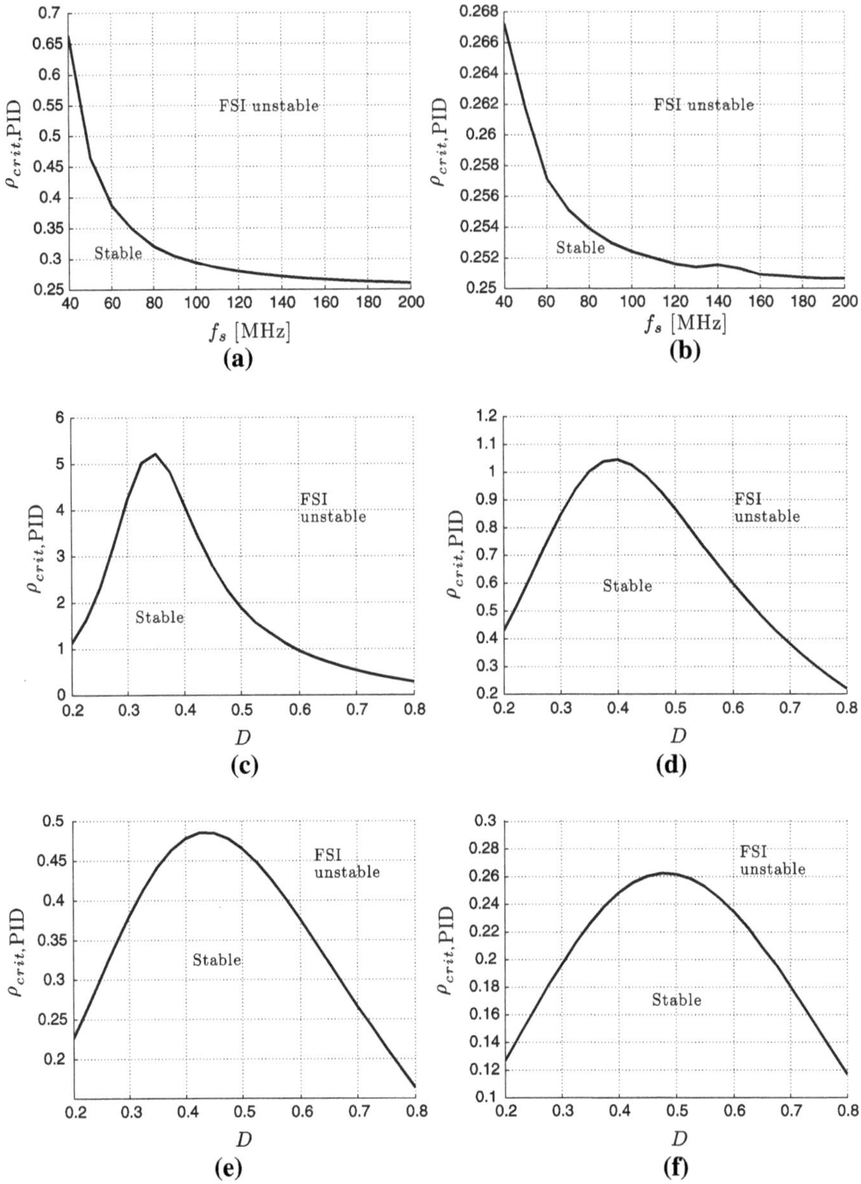

Fig. 3.25 $\rho_{crit,\text{PID}}$ as a function of the PID pole ω_{p2} and **a–b** the switching frequency f_s and **c–f** the duty cycle D. Controller parameters: $\omega_{z1} = 100$ krad/s, $\omega_{z2} = 1$ Mrad/s with $V_{ref} = 3$ V and $\omega_s = 2\pi f_s = 314$ Mrad/s and other parameters are the same as in the CMC case

This section has addressed the prediction of the FSI boundary in a full-state feedback (voltage and current loops) buck regulator providing a simplified expression that joins both control loops. The analysis of this expression allows to know the effect of each parameter upon the stability boundary. However, the main limitation of such index-based approach is that it lacks a frequency domain representation, which is of strong interest to synthesize new controllers and compensators. The next section will develop a frequency domain model that eventually can lead to understand in a more comprehensive way the FSI phenomena and to synthesize new FSI controllers.

3.8 A Frequency Domain Model for Overall Stability Boundary Prediction

The previous sections have tackled the prediction of the FSI boundaries starting from the ripple-based index approach. As it was mentioned in Sect. 3.5, this approach has the advantage that it is possible to directly derive from it the dependence of the FSI boundary upon circuit parameters hence providing a design-oriented closed-form expression for predicting the boundary of unstable behavior. However, starting from such index, it is not possible to obtain a frequency-domain representation that facilitates the synthesis of new controllers to avoid instability.

This section addresses the prediction of the FSI boundary from a frequency domain standpoint for a buck converter, with the final aim of synthesizing new controllers in subsequent chapters. The frequency domain model will be derived from the discrete-time model and will take advantage of the relationship between the ripple of the control signal and the modulator gain demonstrated in Appendix C.

Starting from the discrete-time model derived in Sect. 3.3, the recursive state-equations in terms of the state variable perturbation $\tilde{\mathbf{x}}_n$ and of the control signal duty cycle perturbation $\tilde{d}_n$ along with their associated small-signal state-matrix $\mathbf{J}_x$, given in Eq. 3.19, and the input vector $\mathbf{J}_d$, given in Eq. 3.20, can be expressed as:

$$\tilde{\mathbf{x}}_{n+1} = \mathbf{J}_{\mathbf{x}}\tilde{\mathbf{x}}_n + \mathbf{J}_d\tilde{d}_n \tag{3.75}$$

Supposing zero initial conditions, the expression can expressed in the z-domain as:

$$z\tilde{\mathbf{x}}_n = \mathbf{J}_{\mathbf{x}}\tilde{\mathbf{x}}_n + \mathbf{J}_d\tilde{d}_n \tag{3.76}$$

Let us define the feedback variable $y_n = \mathbf{C}\Phi_1(DT)\mathbf{x}_n$, with $\mathbf{C}$ being a unit vector depending on which of the state variables is used in the feedback. For instance, for the converter under VMC one has $\mathbf{C}$=[1 0]. Let us also define the small signal feedback variable $\tilde{y}_n = \mathbf{C}\Phi_1(DT)\tilde{\mathbf{x}}_n$. Therefore, the z-domain transfer function is:

$$H_{d,dis}(z) = \frac{\tilde{y}_n}{\tilde{d}_n} = \mathbf{C}\Phi_1(DT)(z\mathbf{I} - \mathbf{J}_{\mathbf{x}})^{-1}\mathbf{J}_d \tag{3.77}$$

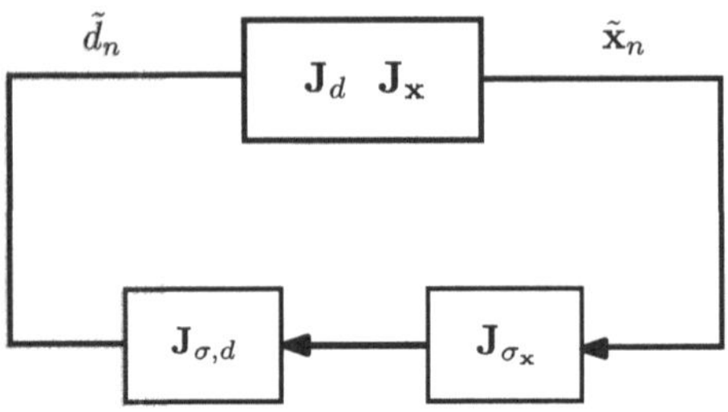

Fig. 3.26 Equivalent block diagram of the VMC converter in the discrete-time domain

Furthermore, from the description of the switching surface σ_d in Eq. 3.17 for a VMC converter, it is possible to calculate the feedback loop and the modulator gains:

$$H_{c,dis}(z) = \frac{\tilde{\sigma}}{\tilde{y}_n} = k_v \tag{3.78}$$

$$H_{m,dis}(z) = \frac{\tilde{d}_n}{\tilde{\sigma}} = \mathbf{J}_{\sigma_d}^{-1} = (k_v \mathbf{C}\dot{\mathbf{x}}(DT^-) + m_c)^{-1} f_s \tag{3.79}$$

where k_v is the gain of the feedback voltage. Finally, the complete loop transfer function of the system $T_{dis}(z)$, shown in Fig. 3.26, can be expressed as a function of the controller and modulator gains:

$$T_{dis}(z) = H_{m,dis}(z)H_{c,dis}(z)H_{d,dis}(z) = \mathbf{J}_{\sigma_d}^{-1}\mathbf{J}_{\sigma_\mathbf{x}}(z\mathbf{I} - \mathbf{J_x})^{-1}\mathbf{J}_d \tag{3.80}$$

The term $\mathbf{J}_{\sigma_\mathbf{x}} = -\mathbf{K}\Phi_1(DT) = k_v\mathbf{C}\Phi_1(DT)$ is given in Eq. 3.22. The poles of the closed loop system, from which it is possible to carry out a stability analysis, can be obtained from:

$$1 + T_{dis}(z) = 0 \tag{3.81}$$

Note that this equation is equivalent to the characteristic polynomial equation of the Jacobian matrix given in Eq. 3.27.

The loop transfer function in a VMC with proportional feedback can be easily defined according to the converter $H_{d,dis}(z)$, the controller $H_{c,dis}(z)$ and the modulator $H_{m,dis}(z)$ transfer functions:

$$H_{d,dis}(z) = \mathbf{C}\Phi_1(DT)(z\mathbf{I} - \mathbf{J_x})^{-1}\mathbf{J}_d \tag{3.82}$$

$$H_{c,dis}(z) = k_p \tag{3.83}$$

$$H_{m,dis}(z) = \left(k_p\dot{v}_C(DT^-) + V_m f_s\right)^{-1} f_s \tag{3.84}$$

Note that $H_{d,dis}(z)$ depends upon all converter parameters, but it is possible to demonstrate that, for a buck converter, it does no depend upon the duty cycle D:

$$\begin{aligned} H_{d,dis}(z) &= \mathbf{C}\Phi_1(DT)[(z\mathbf{I} - \Phi(T))^{-1}\Phi_2(\overline{D}T)]\Delta\dot{\mathbf{x}}T \\ &= \mathbf{C}\Phi_1(DT)\frac{\text{adj}(z\mathbf{I} - \Phi(T))}{\det(z\mathbf{I} - \Phi(T))}\Phi_2(\overline{D}T)\Delta\dot{\mathbf{x}}T \\ &= \mathbf{C}\frac{\Phi(T)}{\det(z\mathbf{I} - \Phi(T))}\Delta\dot{\mathbf{x}}T \end{aligned} \tag{3.85}$$

being $\Phi(T) = \Phi_1(DT)\Phi_2(\overline{D}T) = \Phi_1(DT)\,\text{adj}(z\mathbf{I} - \Phi(T))\Phi_2(\overline{D}T)$. On the other hand, the controller and the modulator transfer functions are constant and do not depend upon the z variable.

3.8.1 *Discrete-Time Model: From z-Domain to Frequency Domain Representation*

The z-domain model derived in the previous section has been also called sampled-data model in the literature (Lau, 1987), (Tymerski, 1993), (Verghese et al., 1986).

Having obtained the z-domain transfer function of the buck switching power converter $T_{dis}(z)$, its frequency response $T_{dis}(e^{j\omega T})$ can be easily derived by substituting, $z = e^{j\omega T}$ (Oppenheim et al., 1999), therefore obtaining:

$$T_{dis}(e^{j\omega T}) = H_{m,dis}(e^{j\omega T})H_{c,dis}(e^{j\omega T})H_{d,dis}(e^{j\omega T}) \tag{3.86}$$

Note that this frequency response considers a zero order hold effect. This transformation has been used in (Ridley, 1989) for a CMC case since it can be demonstrated that, in such a case, the PWM action can be approximated by a sample-and-hold, although this statement is not valid for the case of VMC. Alternative z-domain to frequency domain transformations are explored in (Sun et al., 2000b).

In the case of $H_{d,dis}(e^{j\omega T})$, the transfer function can easily be obtained from the z-domain by means of the aforementioned transformation. The complete feedback transfer function in the frequency domain will be completed by the controller frequency response $H_{c,dis}(e^{j\omega T}) = k_p$ and the modulator gain $H_{m,dis}(e^{j\omega T})$, which can be considered as a simple constant gain in the frequency domain. As it was demonstrated in Appendix C, the output voltage ripple $\hat{V}_C$ is proportional to its derivative at the switching instant, so that:

$$H_{m,dis}(e^{j\omega T}) = \frac{1}{4k_p\hat{V}_C + V_m} \tag{3.87}$$

The final frequency domain response $T_{dis}(e^{j\omega T})$ is depicted in Fig. 3.27, and the FSI condition in the frequency domain can be expressed as:

$$|T_{dis}(e^{j\omega_s T/2})| = |T_{dis}(e^{j\pi})| = |T_{dis}(-1)| = 1 \tag{3.88}$$

$$\angle T_{dis}(e^{j\omega_s T/2}) = \angle T_{dis}(e^{j\pi}) = \angle T_{dis}(-1) = -180^\circ \tag{3.89}$$

The validation of the frequency domain representation and FSI prediction is shown in Fig. 3.27 for $V_g = 6\,\text{V}$, $V_{ref} = 3\,\text{V}$, $R = 2.5\ \Omega$, $L = 66\,\text{nH}$, $C = 20\,\text{nF}$, $f_s = 50\,\text{MHz}$, $V_m = 1\,\text{V}$, $k_p = 3$ and $\omega_{z1} = 1\,\text{Mrad/s}$. It can be observed that for the set of parameter values for which $T_{dis}(-1) = -1$, one of the eigenvalues of the Jacobain matrix is equal to -1.

3.8.2 The Averaged Model and the Discrete-Time Model Frequency Domain Discrepancies

This section shows in a frequency domain representation the discrepancies between the averaged model, which has been discussed in Sect. 3.1, and the discrete-time model. While in both models the controller is assumed constant and equal to $G_c(j\omega) = H_{c,dis}(e^{j\omega T}) = k_p$ there are discrepancies regarding to the modulator and output-to-control transfer function. The difference between converter control-to-output transfer functions in the averaged and the discrete-time model, $H_{d,avg}(j\omega)$ and $H_{d,dis}(e^{j\omega T})$ respectively, is shown in Fig. 3.28.

From this figure, it can be observed that at low frequency both transfer functions are consistent, while they differ close to half of the switching frequency due to the switching action being eliminated in the averaged model.

Despite the difference within the high-frequency band, our interest is just centered in the frequency response at half of the switching frequency due to its implication in the prediction of the FSI boundary, as it was demostrated in the previous section. The difference between both models at half of the switching frequency are quantified in Fig. 3.29. Note that the error is almost constant in terms of the design-space parameters Γ and τ. Therefore, the maximum error between both models at half of the switching frequency can be approximated by 8 dB, being this quantity a conservative limit.

The exploration of the design-space as a function of the duty cycle makes no sense since neither the averaged model $H_{d,avg}e^{(j\omega T)}$ nor the discrete-time model $H_{d,dis}(e^{j\omega T})$ depend on this parameter, as it was demonstrated in Eq. 3.85.

Regarding the modulator transfer functions, note that both models have a constant value, but have different expressions. Then, using the discrete-time model modulator gain $H_{m,z}(e^{j\omega T})$ into the averaged model will imply a distortion of the SSI boundary prediction. Therefore the appropriate modeling of the modulator is a key element for the frequency model to be able to predict the overall stability boundary in the frequency domain.

Fig. 3.27 Representation of $T_{dis}(e^{j\omega T})$ and the validation of the prediction of FSI boundary with discrete-time model. VMC buck converter with $k_p = 4.3$, being in the boundary of FSI, and obtaining $|T_{dis}(e^{j\omega_s T/2}|) \approx 1$ and the system Jacobian eigenvalues $(-0.9923 - 0.6755)$

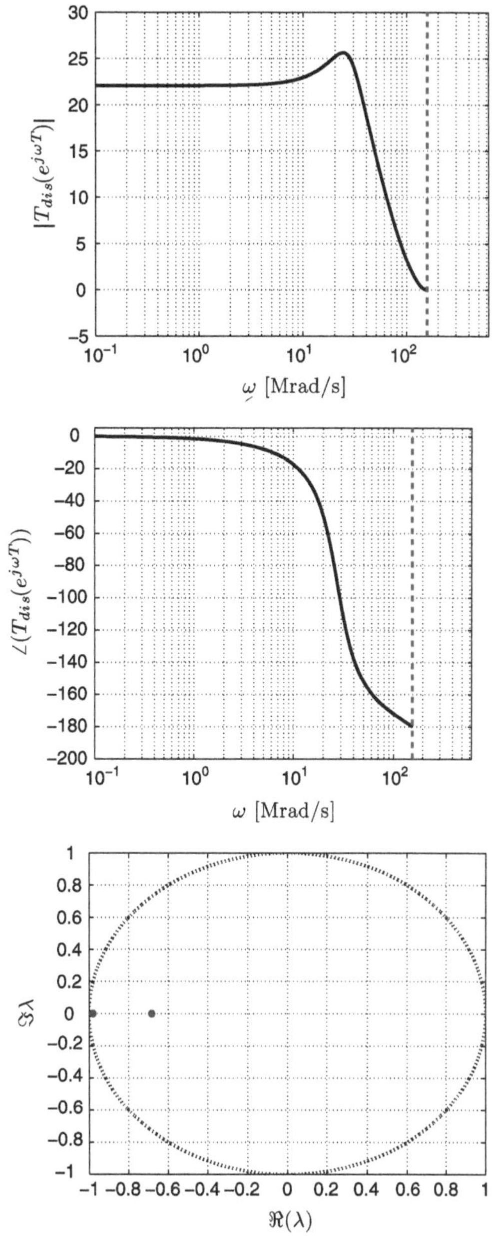

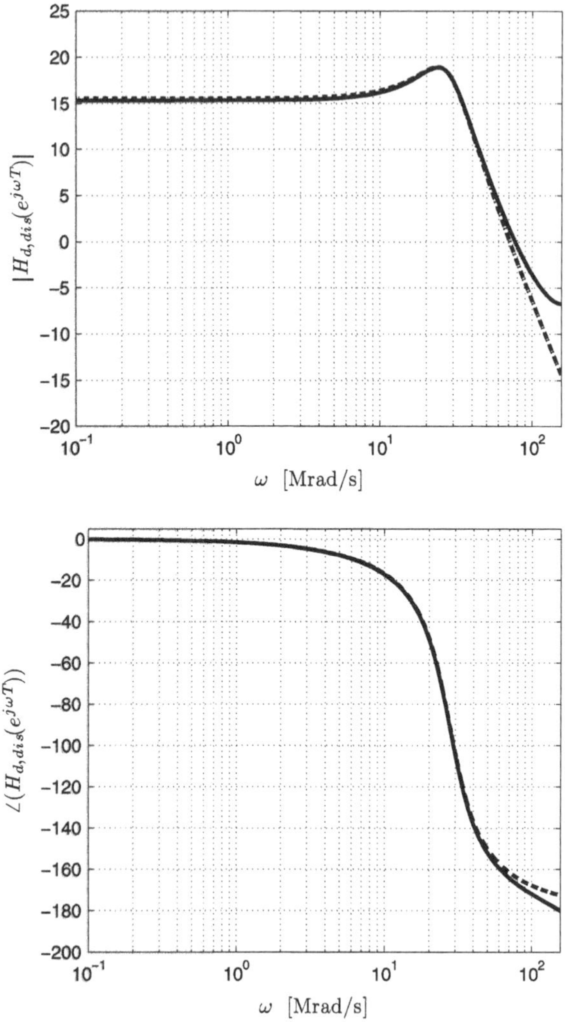

Fig. 3.28 Bode plots of $H_{d,avg}(j\omega)$(*dashed*) given in Eq. 3.2 and $H_{d,dis}(e^{j\omega T})$ (*solid*) given in Eq. 3.82

3.8.3 The Modulator Frequency Response

While in the discrete-time model the modulator gain is just a constant, including the voltage ripple and the ramp slope, this simple expression neglected the real nature of the modulator, since, indeed, the modulator gain has a dependency on the control signal waveform $v_{con}(t)$. Note that the modulator transfer function is obtained by sampling the derivative of the control signal, whose period is the same as the discrete-time model sampling period hence being a constant from the discrete-time model standpoint. However, any controller that affects the control signal will affect this gain, hence it can be considered that there is an associated frequency response to

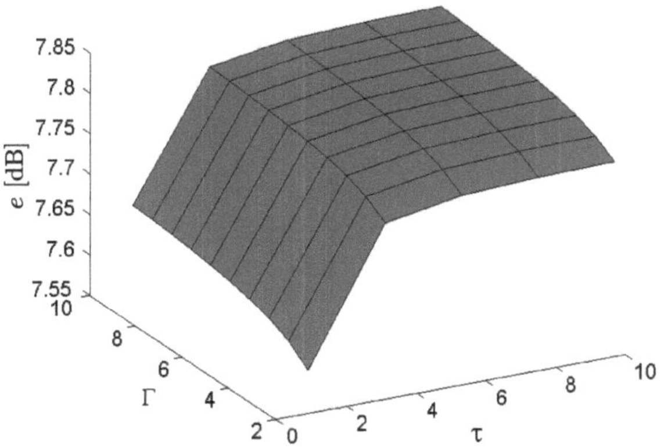

Fig. 3.29 Error (in dB) in the magnitude between $H_{d,avg}(j\omega_s/2)$ and $H_{d,dis}(e^{j\omega_s T/2})$

such modulator. Therefore, it must depend upon the other system transfer functions, namely the converter and the controller. There are some works dealing with obtaining the small-signal PWM modulator model such as (Tymerski and Li, 1993; Vorperian, 1990), but they are mainly based on complex mathematical analysis. With the aim of not losing the design-oriented standpoint, this section proposes a simple frequency domain model for the modulator starting from the observation that the feedback signal slope at the switching instant $\dot{v}_C(DT^-)$ is proportional to the converter output ripple amplitude $\hat{V}_C$ for the case of a VMC buck converter.

While the ripple amplitude $\hat{V}_C$ of the output voltage is just a measure of the time-domain output voltage waveform $v_C(t)$, it has an important advantage since it is highly correlated with the first harmonic of the ripple waveform $v_c(t)$. This can be demonstrated from the Fourier series of the square-wave diode voltage $v_d(t)$ with the duty cycle D:

$$v_d(t) = a_0 + \sum_{n=1}^{\infty} a_n \cos(n\omega_s t) + \sum_{n=1}^{\infty} b_n \sin(n\omega_s t) \tag{3.90}$$

where

$$a_0 = \frac{1}{T}\int_0^T V_{gq(t)} = DV_g \tag{3.91}$$

$$a_n = \frac{2}{T}\int_0^T V_{gq(t)} \cos(n\omega_s t) = \frac{V_g}{\pi n}\sin(2\pi n D) \tag{3.92}$$

$$b_n = \frac{2}{T}\int_0^T V_{gq(t)} \sin(n\omega_s t) = \frac{V_g}{\pi n}(1 - \cos(2\pi n D)) \tag{3.93}$$

The output voltage ripple waveform $v_C(t)$ can be expressed as a function of the PWM Fourier series expression and the frequency response of the buck converter

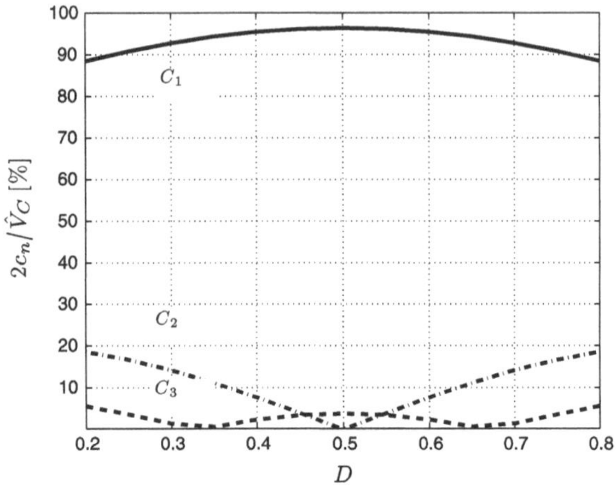

Fig. 3.30 Comparison in % for the output waveform harmonic amplitude c_n: c_1 (*solid*) c_2 (*dash-dot*) and c_2 (*dots*) with output ripple measure $\hat{V}_c$

low-pass filter G_p.

$$v_C(t) = c_o + \sum_{n=1}^{\infty} c_n \cos(n\omega_s t + \varphi_n) \tag{3.94}$$

where $c_n = A_n|G_n|$, $G_n{=}G_p(jn\omega_s)$ is the attenuation of the LC filter at the $n-$th harmonic of the switching frequency, $A_n = \sqrt{a_n^2 + b_n^2}$ and $\varphi_n = -\arctan(b_n/a_n)$. Figure 3.30 compares the amplitude of each harmonic c_n with the total ripple magnitude $\hat{V}_C/2$ (peak), showing that both measures are very coincident (over 90 %). This implies that the ripple magnitude can be approximated by the first harmonic level due to the intrinsic low-pass filter of the converter, and hence any effect (attenuation or amplification) on such harmonic c_1 can be directly applied on the ripple magnitude $\hat{V}_C$. Therefore, the output voltage ripple $\tilde{v}_c(t)$ can be expressed as:

$$\tilde{v}_C(t) \approx A_1|G_1|\cos(\omega_s t + \varphi_1) \approx \frac{\hat{V}_C}{2}\cos(\omega_s t + \varphi_1) \tag{3.95}$$

where

$$A_1 = \frac{2V_g}{\pi}\sin(\pi D) \tag{3.96}$$

Then, the associated frequency domain representation can be expressed as:

$$\tilde{v}_C(\omega) \approx A_1|G_1|\delta(\omega - \omega_s) \approx \frac{\hat{V}_C}{2}\delta(\omega - \omega_s) \tag{3.97}$$

where $\delta()$ is the Dirac delta impulse function. Having demonstrated that the ripple magnitude is highly correlated with the first harmonic amplitude of the control signal, and that such harmonic can be easily represented in a frequency domain, a new small-

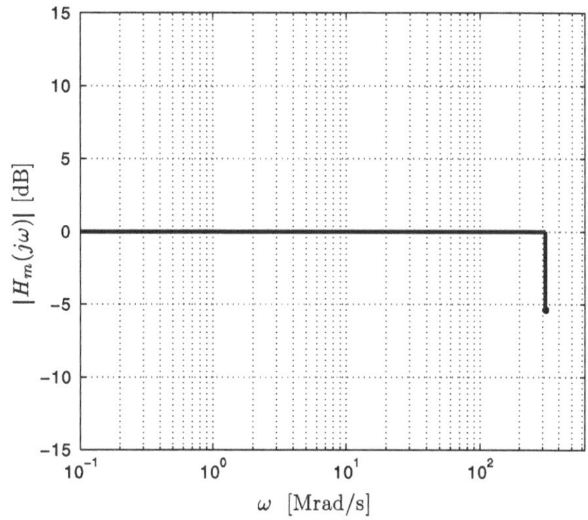

Fig. 3.31 Modulator transfer function $|H_{mf}(j\omega)|$ with constant gain in all the frequency range below f_s ($V_m = 1$ V) and a tone at the switching frequency

signal modulator frequency response can be constructed:

$$H_{mf}(j\omega) = \begin{cases} \dfrac{1}{V_m} & \text{if } \omega \neq \omega_s \\ \dfrac{1}{4k_p\hat{V}_C + V_m} & \text{if } \omega = \omega_s \end{cases}$$

The new small-signal modulator transfer function frequency response depends upon the control signal, through the ripple magnitude at the switching frequency. This is in agreement with previous studies that demonstrated that averaged model modulator gain, namely $1/V_m$, is valid up to one half of the switching frequency.

The modulator frequency response magnitude is illustrated in Fig. 3.31, showing that it is constant for the all of frequencies except at the switching frequency where there is a tone the amplitude of which is proportional to the ripple magnitude. Therefore, only the effect of a compensator $G_c(j\omega)$ upon such transfer function at the switching frequency is of interest.

3.8.4 Extended Discrete-Time Model

The modulator small-signal transfer function frequency response developed in the previous section can be applied in both discrete or averaged models. In the first case, the feedback transfer function $T_{dis}(e^{j\omega T})$ can be modified as follows

$$T_{disf}(e^{j\omega T}) = H_c(e^{j\omega T})H_{d,dis}(e^{j\omega T})H_{mf}(j\omega) \tag{3.98}$$

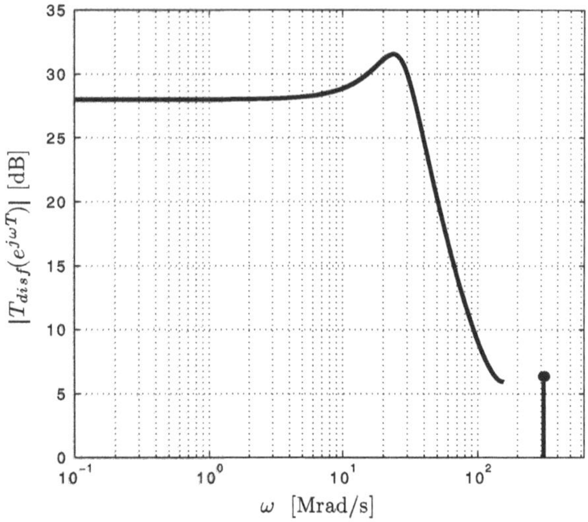

Fig. 3.32 a Frequency response of $T_{disf}(e^{j\omega T})$, in this representative example the system being at the boundary of FSI conditions, with $k_p = 4.31$ and $SM_{FS} = 0.01$ dB

Note that the new stability criterion, instead of just ensuring that the gain at one half the switching frequency is lower than 0 dB, as it was established in Eq. 3.81, will imply to compare the loop response at two frequencies. In particular, the magnitude of $T_{disf}(e^{j\omega T})$ at half of the switching frequency must be lower than its magnitude at the switching frequency to ensure stability. Therefore, it is possible to define a new FSI stability margin (SM_{FS}) as follows:

$$SM_{FS} = 20\log(|1 + \frac{4G_c(\omega_s)|}{V_m}\hat{V}_C) - 20\log(|T_{disf}(e^{j\omega_s T/2})|) \qquad (3.99)$$

Therefore, FSI will be avoided if $SM_{FS} > 0$. Note that the frequency responses $T_{disf}(e^{j\omega T})$ and $H_c(e^{j\omega T})H_{d,dis}(e^{j\omega T})/V_m$ are equals up to one half of the switching frequency. Both frequency responses are plotted in Fig. 3.32.

In Fig. 3.33 the stability boundaries obtained from the exact discrete-time model and by using the FSI stability margin given in Eq. 3.99 from the frequency domain model are depicted showing a good agreement, thereby demonstrating the accuracy of the new model. Furthermore, note that the stability boundary as a function of the duty cycle is not a perfect parabola (in agreement with what was observed for the ripple-based index approach) since $H_{mf}(j\omega)$ has an independent term at the switching frequency that does not depend upon any system parameter hence the stability boundary is not directly proportional to the ripple magnitude.

Note that these results are equivalent to the ones obtained in Sect. 3.3.4 using the ripple within the discrete-time model for predicting FSI boundary (since both include the same kind of assumptions).

Whereas the stability condition in Eq. (3.99) implies that the ripple has to be kept within a certain threshold to ensure stability to avoid FSI, according to the ripple-based index approach, developed in Sect. 3.2, the input modulator ripple has

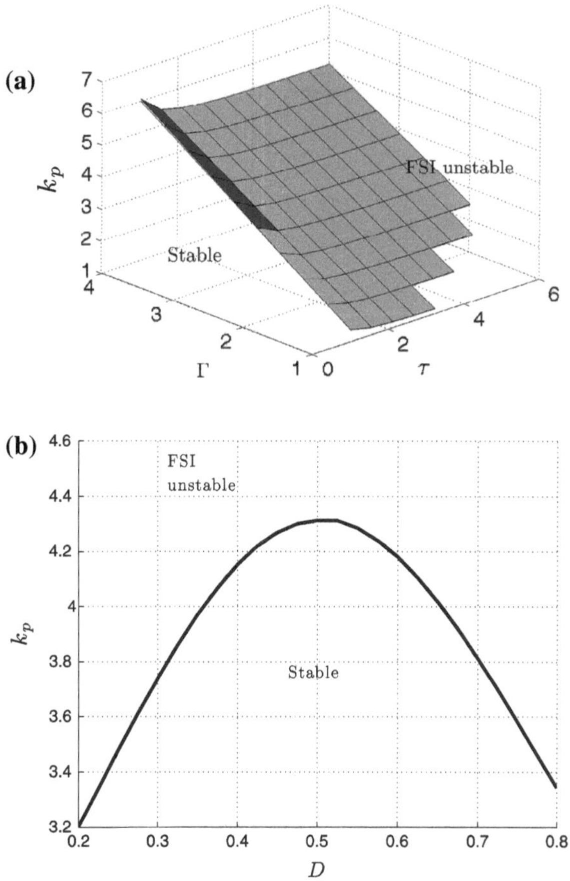

Fig. 3.33 Stability boundary obtained from the discrete-time model and from the FSI condition given in Eq. 3.99 **a** as a function of parameters Γ and τ **b** as a function of the duty cycle D

to be kept under a certain level to avoid the exhibition of the same instability. In principle this seems a contradiction but the key point is to analyze the procedure that has been carried out to obtain the ripple-based index stability condition from the discrete-time model. Note that in Eq. 3.39, both sides of the analytical equation depend upon the same the system parameters, hence a common factor was isolated and then transformed into the ripple-based index. On the other hand, the frequency model represents independently both sides of the same equation. Indeed, the ripple-based index approach captures, indirectly, and expresses in a simple condition, the condition for which the harmonic at the switching frequency is equal to the harmonic at half of the switching frequency.

Whereas the ripple-based index has a clear usefulness for prediction of the FSI boundary, the frequency domain has the clear aim of facilitating the synthesis of new controllers. From such frequency domain representation, it is possible to derive that

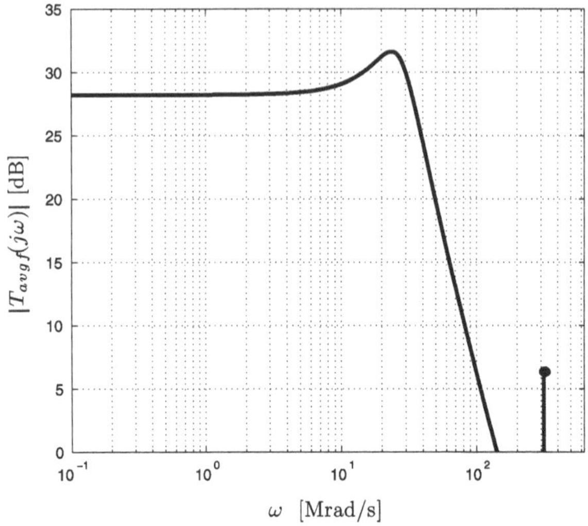

Fig. 3.34 **a** Frequency domain representation of $T_{avgf}(j\omega)$ at the boundary of FSI conditions with $k_p = 4.31$, with $SM_{FS} = 7.8\,\mathrm{dB}$

the control of such instabilities can be tackled either by attenuating the harmonic at half of the switching frequency standpoint or by amplifying the switching frequency harmonic. New FSI controller synthesis will be addressed in the next chapter taking advantage of this observation.

3.8.5 Complete Design-Oriented Frequency Domain Model

Although the modulator transfer function based on the ripple representation in the frequency domain has a clear design-oriented purpose, the model given in the previous section still requires to develop the discrete-time model to obtain $H_{d,dis}(e^{j\omega T})$, which lacks a bottom-up circuit derivation standpoint.

The design-oriented standpoint of the averaged model along with the simplicity of the modulator transfer function $H_{mf}(j\omega)$, which depends only upon the ripple, facilitates the derivation of a new design-oriented small signal complete loop transfer function $T_{avgf}(j\omega)$ for predicting both SSI and FSI.

$$T_{avgf}(j\omega) = H_{d,avg}(j\omega)G_c(j\omega)H_{mf}(j\omega) \tag{3.100}$$

The magnitude frequency response of $T_{avgf}(j\omega)$ is shown in Fig. 3.34. Taking into account that in Sect. 3.8.2 it has been demonstrated that the difference between the control-to-output transfer function in the discrete-time model and the averaged model at half of the switching frequency is of a maximum of 8 dB, it is possible to

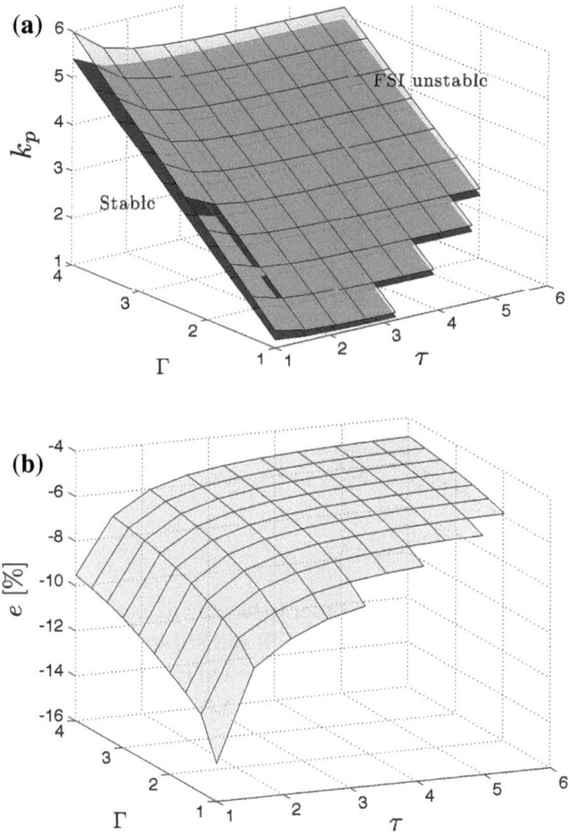

Fig. 3.35 **a** FSI boundary surface obtained from the discrete-time model (*black*) and from the frequency domain model stability condition given in Eq. 3.101 (*white*). **b** Error between both surfaces in %

rewrite the stability margin given in Eq. 3.99 as follows

$$SM_{FS} = 20\log(T_{avgf}(j\omega_s) - 20\log(|T_{avgf}(j\omega_s/2)|) > 8\,\text{dB} \tag{3.101}$$

In Fig. 3.35 the FSI boundary, obtained from the discrete-time model and from the stability condition given in Eq. (3.101), is shown. It can be observed that the new model along with the stability condition gives a conservative boundary for predicting FSI.

3.8.6 Extension to a PI Compensator

In the previous section, an approximated design-oriented circuit-based model for a buck converter under a proportional control has been obtained. This section extends

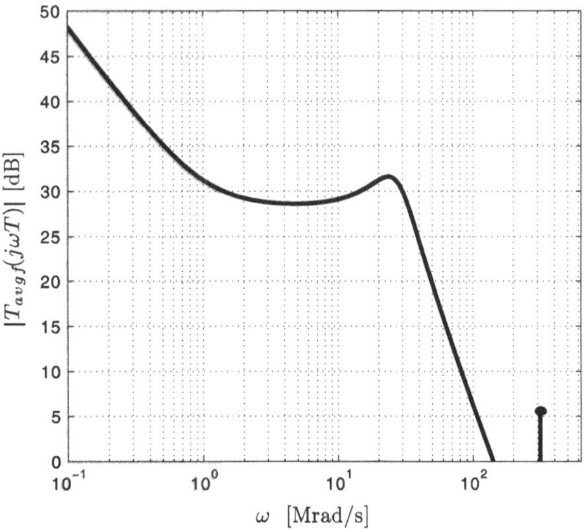

Fig. 3.36 Frequency domain magnitude response of $T_{avgf}(j\omega)$ obtained by joining the averaged model of a VMC buck converter, under a PI compensator, along with the modulator transfer function $H_{mf}(j\omega)$

the approach to a PI compensator in order to be able to explore the usefulness of the approach for concurrently predicting both FSI and SSI stability boundaries since with a proportional controller only FSI can be exhibited in a buck converter under VMC. The development of the discrete-time model of a VMC buck converter under a PI compensator, requires to proceed as in the case of a proportional control, increasing the number of state variables and adjusting properly the gain vector (see Appendix A.1) hence implying a certain complexity. However, an alternative approach can be obtained from the previous model, by using the averaged model along with the ripple-based modulator transfer function. The PI compensator frequency response is:

$$G_c(j\omega) = k_p \frac{j\omega + \omega_{z1}}{j\omega} \tag{3.102}$$

The total loop transfer function $T_{avgf}(j\omega)$ is shown in Fig. 3.36. In this case, two conditions have to be satisfied: on the one hand, the Nyquist stability condition (slow scale), and on the other hand, the stability condition given in Eq. 3.101 (fast scale). Figure 3.37 shows the stability boundary obtained from the discrete-time model and from the complete design-oriented frequency domain model, by sweeping the proportional gain k_p and the PI zero ω_{z1}. The model is able to accurately predict the SSI stability boundary, since it directly includes the averaged model, and gives an accurate conservative limit for predicting the FSI stability boundary, hence demonstrating the benefits of this combined approach for predicting both FSI and SSI .

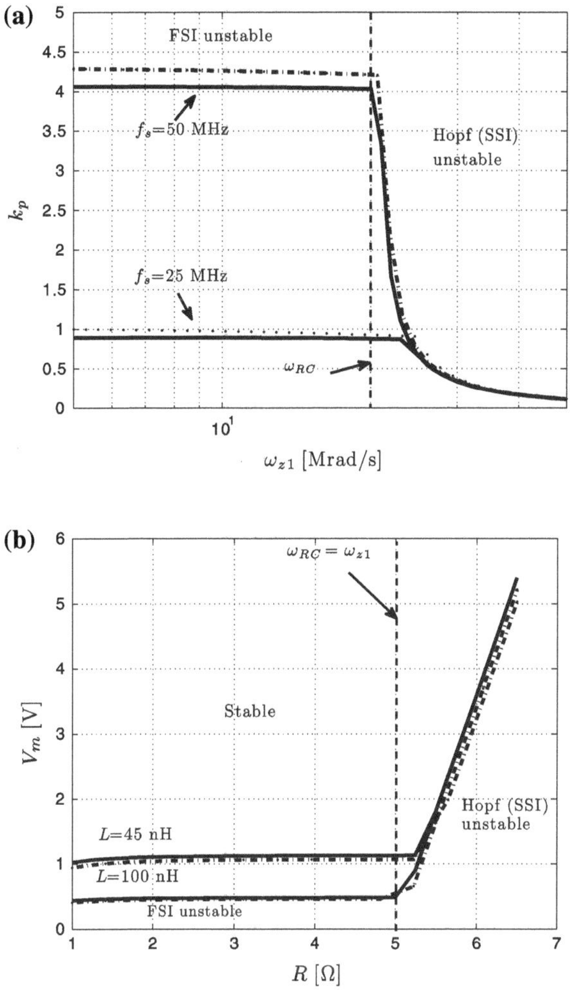

Fig. 3.37 Overall stability boundary obtained from the discrete-time model (*dash-dot*) and results from the frequency domain model (*solid*), using the Nyquist stability criterion (*phase and gain margins*) along with the FSI stability condition given in Eq. 3.101, over the design parameter space **a** ω_{z1} and k_p

3.9 Concluding Remarks

This chapter has presented an approach to predict the occurrence of FSI based on the ripple magnitude. The ripple-based index allows compressing different nature parameters into a single index hence facilitating the quantitative prediction of the FSI occurrence considering the complete parameter design-space. The accuracy of this approach has been contrasted with the results obtained from the consolidated discrete-time models hence demonstrating its validity. Furthermore, the chapter has demonstrated that, by using a small-signal model for the modulator, the ripple component explicitly appears within such discrete-time model hence allowing to derive a closed-form expression that mathematically validates the ripple-index approach. The approach has also been validated for DCM. Furthermore, the approach has been also extended to CMC obtaining a simple stability condition that allows predicting the stability boundary taking into account the outer voltage feedback loop and the external ramp slope. The work has also validated the general applicability of the CMC stability condition for a PID controller case. Finally, the chapter has proposed a new modulator frequency response that, along with the averaged model, is able to predict the overall system stability boundaries. Beyond the fact of having a single model for concurrently predicting both kinds of instability, the obtained frequency representation allows to address the synthesis of new controllers. The next chapter will revisit FSI controllers proposed up to now and will improve their performance elaborating from the results obtained from the frequency model proposed in this chapter.

Chapter 4
Control of FSI in Switching Power Converters

4.1 Introduction

The previous chapter has addressed the prediction of FSI in a VMC buck converter, directly relating it to the ripple magnitude at the modulator input. Beyond the fact of being able to quantitatively predict such stability boundary from a ripple-based approach, the chapter puts forward the fact that the design trends towards miniaturization, namely low area and low relative switching frequency, which directly implies an increase upon ripple, is indirectly limited by the exhibition of FSI. With this interest in sight, this chapter tackles the synthesis of FSI (chaos) controllers from an engineering standpoint, hence facilitating their design and implementation and understanding their impact upon overall stability.

This section considers a VMC buck converter with a PI compensator along with additional circuitry, namely an FSI controller $G_{FS}(s)$ to avoid the exhibition of FSI, as it is shown in Fig. 4.1. The nominal parameters used in this chapter are the same as in the previous one: $V_g = 6$, $V_{ref} = 3\,\mathrm{V}$, $f_s = 50\,\mathrm{MHz}$, $V_m = 1\,\mathrm{V}$, $L = 66\,\mathrm{nH}$, $C = 20\,\mathrm{nF}$, $R = 2.5\,\Omega$, $k_p = 3$ and $\omega_{z1} = 1\,\mathrm{Mrad/s}$. The bifurcation diagram by sweeping k_p of such converter, without any FSI controller is illustrated in Fig. 4.2.

The starting point will be revisiting the existing chaos controllers. Their main drawbacks are, on the one hand, the fact that they are mainly analyzed from a mathematical standpoint (Lyapunov exponent) (Pyragas 1992, 1995; Batlle et al. 1997) but avoiding an engineering standpoint that facilitates their design.

The chapter analyzes different chaos controllers from a frequency domain standpoint which allows to understand their principle of operation and gives way to propose alternative implementation-aware approaches. Subsequently, these controllers are extended focusing on facilitating miniaturization, thereby rejecting instabilities but at the same time also improving the power-related performance metrics, namely area, efficiency and ripple.

Finally, the chapter ends up by comparing the stability margins obtained by the different controllers, its capability to extend the power-related performance metrics and their impact upon dynamics metrics such as the system transients.

E. Rodríguez Vilamitjana et al., *Chaos in Switching Converters for Power Management*, DOI: 10.1007/978-1-4614-2128-3_4,

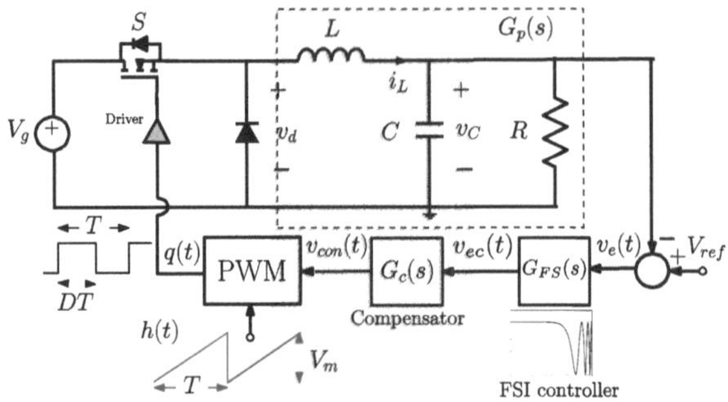

Fig. 4.1 VMC buck converter with a compensator $G_c(s)$ along with an FSI controller G_{FS}(s)

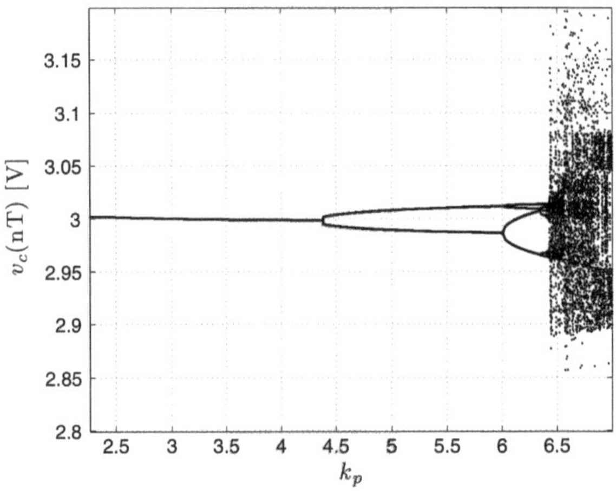

Fig. 4.2 Bifurcation diagram by increasing k_p in a VMC buck converter with a PI compensator showing the route to chaos via period-doubling. $k_{p,crit} = 4.3$

4.2 Time-Delay-Based Chaos Controllers

This section revisits time-delay feedback controllers in order to describe their operation principle and the effect of their parameters upon stability. One of the most studied chaos controller categories in the literature is the Time-Delay Feedback Controller (TDFC) and its extension (ETDFC) (Batlle et al. 1997; Angulo et al. 2007), whose block diagrams are shown in Fig. 4.3.

From the controller structure, the generic TDFC Laplace-domain transfer function can be expressed as:

$$G_{FS}(s) = 1 - \gamma \frac{e^{-sT} - 1}{e^{-sT} - \beta} \tag{4.1}$$

where β is zero for the TDFC and it takes a non-zero value in the case of ETDFC.

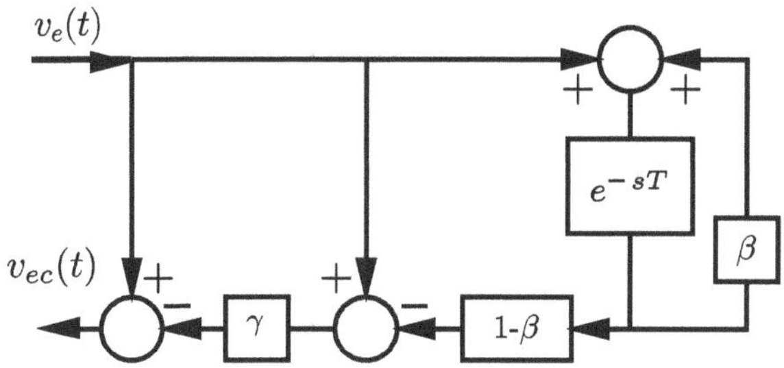

Fig. 4.3 Time-delay feedback control scheme

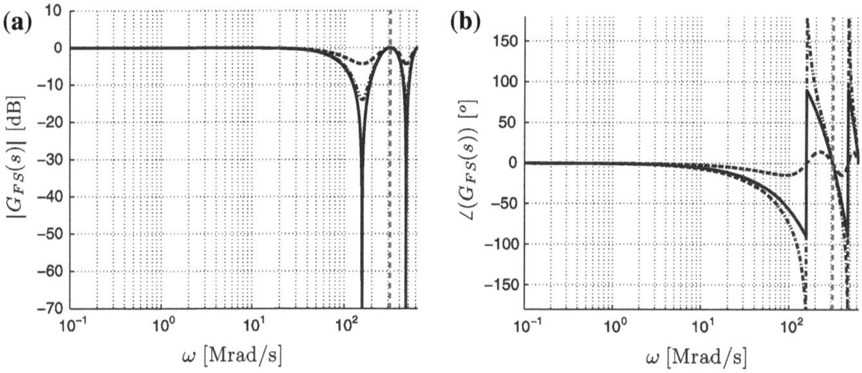

Fig. 4.4 Bode diagram representation (magnitude and phase) of the TDFC with $\gamma = 0.2$ (*dashed*) $\gamma = 0.5$ (*solid*) and $\gamma = 0.6$ (*dot-dashed*) showing the switching frequency (*vertical dashed line*)

4.2.1 The Time-Delay Feedback Controller

This section explores the frequency response of the TDFC, obtained from particularizing the transfer function given in Eq. (4.1) for $\beta = 0$.

The frequency-domain representation depicted in Fig. 4.4, for different values of the parameter γ, shows that the controller provides a frequency-selective comb-filter which attenuates the harmonics at half of the switching frequency and its integer multiples. Note that the controller has a *non-invasive* nature since it does not modify neither the DC component, nor the harmonic at the switching frequency. As it is possible to observe, by increasing γ the attenuation increases (only for γ up to 0.5), and adds an additional phase lag.

The effect of γ upon exhibition of FSI is shown in Fig. 4.5. This figure shows that the controller leads to improve the FSI boundary (even avoiding the chaotic behaviour), but with some strong limitations within the design-space. Namely, on the one hand, there exists a critical minimum value of γ to avoid FSI exhibition, which depends upon the amount of ripple in the converter, considering the direct relationship between stability and ripple demonstrated in the previous chapter. On

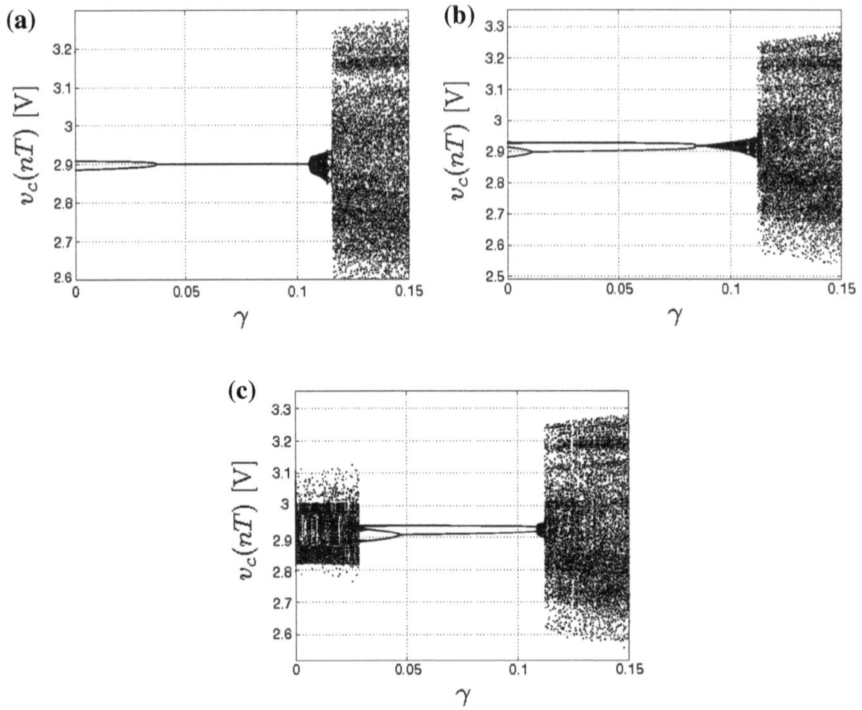

Fig. 4.5 Bifurcation diagram by sweeping γ in a VMC buck converter with TDFC for different values of proportional gain k_p. **a** $k_p = 5$ (in period-doubling without TDFC). **b** $k_p = 6.2$ (in period-four without TDFC). **c** $k_p = 7$ (chaos without TDFC)

the other hand, simulation results show that for a certain critical value of γ (which does not depend upon the ripple) the TDFC yields to exhibit SSI. This is in agreement with the phase response given in Fig. 4.4 for which higher values of γ lead to adding a phase lag to the loop gain, hence making the converter prone to exhibit SSI. This fact, detrimental to SSI, opposes to the controller benefits in terms of FSI hence compromising the overall stability of the system.

Figure 4.6 shows the bifurcation diagram as a function of k_p for different values of γ. As it was pointed out before, by using a too much high value of γ, SSI is exhibited thereby limiting the potential advantages of the controller.

4.2.2 Extended Time-Delay Feedback Controller

The Extended Time-Delay Feedback controller (ETFDC) includes an additional feedback inner loop depending upon a new parameter β as it is shown in Fig. 4.3.

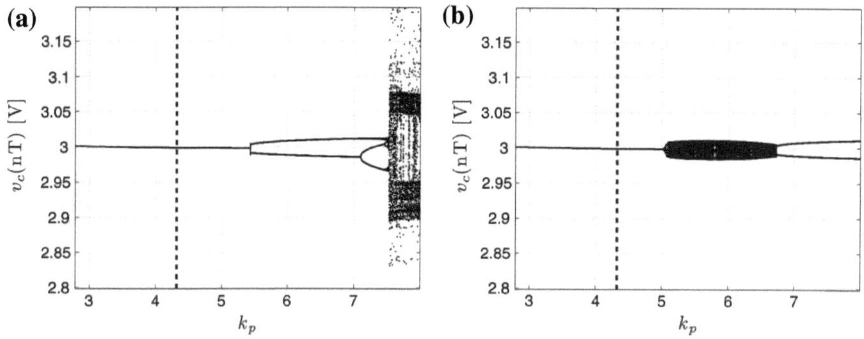

Fig. 4.6 Bifurcation diagram by sweeping the proportional gain k_p in a VMC buck converter with TDFC for different values of γ **a** $\gamma = 0.05$ **b** $\gamma = 0.1$

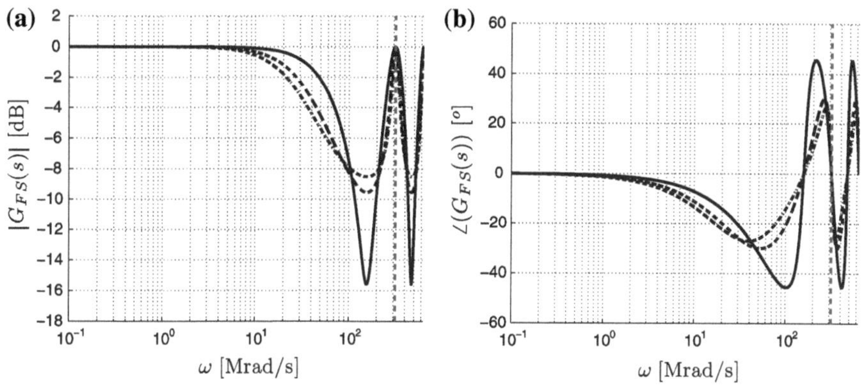

Fig. 4.7 Bode diagram representation (magnitude and phase) of the ETDFC with $\gamma = 0.5$ and $\beta = 0.2$ (*solid*) $\beta = 0.5$ (*dash*) and $\beta = 0.6$ (*dot-dash*) showing the switching frequency (*vertical dashed-line*)

Figure 4.7 shows the effect of such parameter upon its frequency response. By increasing the value of β and keeping γ constant, the phase response is smoothened, but on the other hand the attenuation at half of the switching frequency decreases. In fact, this trade-off allows to use higher values of γ than in the TDFC case, as it is shown in Fig. 4.8, hence improving the FSI boundary, but requiring a higher minimum value of γ to control it.

Figure 4.9 shows the stability boundary surface, obtained from numerical simulations, by sweeping both γ and β parameters in order to be able to guarantee the overall stability margin (considering both FSI and SSI). For low values of β and high values of γ, SSI is exhibited, whereas, as β is increased, the system is more prone to exhibit FSI. The figure also shows the bifurcation diagram for the most optimum stable conditions in terms of FSI, namely $\gamma = 0.2$ and $\beta = 0.5$.

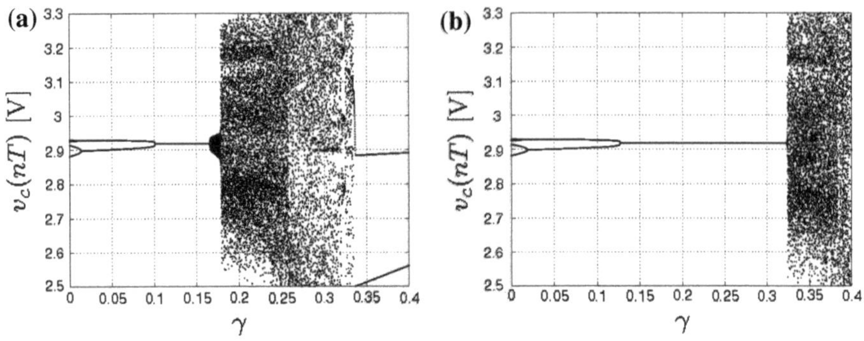

Fig. 4.8 Bifurcation diagram by sweeping γ in a VMC buck converter with ETDFC for different values of β **a** $\beta = 0.2$ and **b** $\beta = 0.5$. $k_p = 6.2$ (period-four without controller)

Regardless of the stability improvement in the ETDFC, both time-delay-based controllers have some important limitation in terms of design and implementation. The first one arises because of the trade-off between SSI and FSI exhibition, which requires a proper selection of γ and β parameters to avoid these instabilities. The lack of a clear criteria to set up these parameters values leads to require a proper exploration as it has been carried out in this section. The second one arises from the implementation difficulties of time-delay-based controllers since both of them are based on a delay module, the implementation of which is challenging in the analog domain.

4.3 Notch-Based Chaos Controllers

The chaos controllers studied in the previous section are based on eliminating the sub-harmonic components from the feedback path to avoid the exhibition of FSI, by means of a delay-based structure, which has implementation difficulties in the analog domain. In Wei-Guo et al. (2010) the time-delay magnitude of the TDFC controller has been replaced by a notch and a high-pass filter in order to simplify its implementability but the structure remains similar, hence being blind to the design-standpoint.

Delay-time-based approaches are based on an additional feedback loop to create a frequency response $G_{FS}(s)$, which provides a comb-like filtering starting with a notch filter at half of the switching frequency. The previous section has characterized that their frequency response can be modified as a function of γ and β parameters.

This section explores the effect of a stop-band filter in the feedback path, the generic transfer function of which is:

$$G_{FS,notch}(s) = \frac{s^2 + 2\xi_1\omega_n + \omega_n^2}{s^2 + 2\xi_2\omega_n + \omega_n^2} \tag{4.2}$$

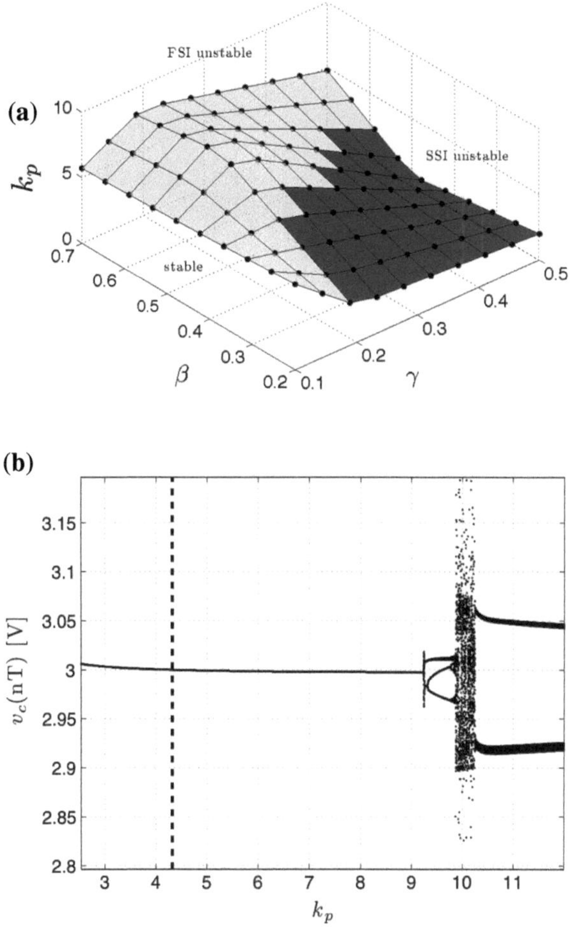

Fig. 4.9 **a** FSI boundary surface, obtained from numerical simulations, in a VMC buck converter with ETDFC as a function of β and γ. FSI (*white*) and SSI (*black*) boundaries are shown. **b** Bifurcation diagram by sweeping k_p with $\gamma = 0.2$ and $\beta = 0.5$

Note that for the case of $\xi_1 = 0$, the transfer function corresponds to a pure notch filter which only depends upon the parameter ξ_2, being the notch resonant frequency ω_n tuned to half of the switching frequency $\omega_n = \omega_s/2$.

The bifurcation diagram in Fig. 4.10 shows that the effect of the parameter ξ_2 leads the converter to exhibit SSI, even for low values of ξ_2 (high quality factor) because of the phase lag. Note that the bifurcation point $k_p = 4.2$ is very close to the one obtained without controller, as it is shown in Fig. 4.2, but in this case exhibiting SSI instead of FSI.

Alternatively and in a similar way as it has been carried out for ETDFC, it is possible to smooth the magnitude and the phase of the transfer function by moving

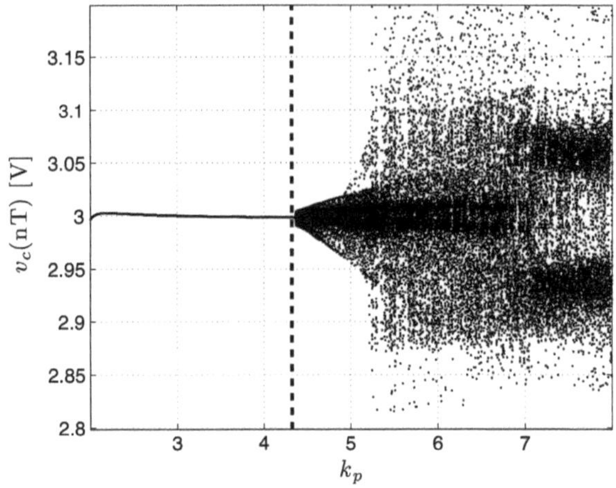

Fig. 4.10 Bifurcation diagram by sweeping the proportional gain k_p in a VMC buck converter with a pure notch filter $\xi_1 = 0$, tuned at half of the switching frequency with $\xi_2 = 0.001$

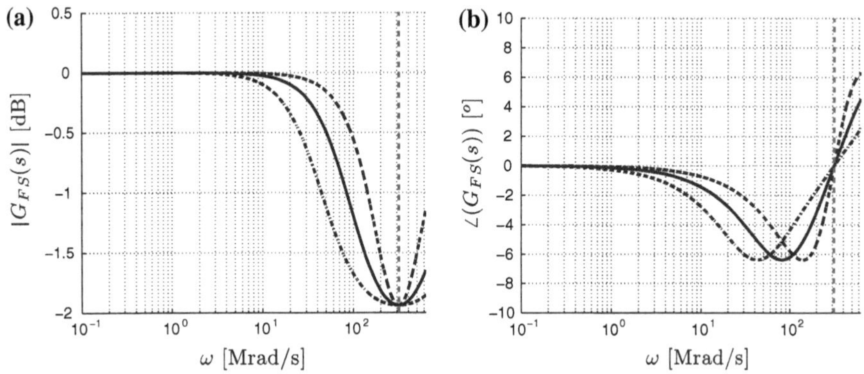

Fig. 4.11 Bode diagram representation (magnitude and phase) of the stop-band controller as a function of ξ_2 keeping constant the attenuation $\Delta_\xi = 0.8$ with $\xi_2 = 1$ (*dash-point*), $\xi_2 = 2$ (*solid*) and $\xi_2 = 4$ (*dash*)

the zeros of the expression in Eq. (4.2) (for $\xi_1 \neq 0$). The transfer function in the frequency domain is shown in Fig. 4.11, in which the design-space can be described as a function of ξ_2 and the attenuation at half of the switching frequency $\Delta_\xi = \xi_1/\xi_2$ (Δ_ξ <1).

The exploration of the FSI stability surface in Fig. 4.12, obtained from the discrete-time model given in Appendix A, unveils similar results as in the ETDFC case, this is, the FSI boundary depends upon the attenuation Δ_ξ and narrowness of the stop-band. The higher the attenuation is, the better stability is obtained in terms of FSI, but, on the other hand, the system is more prone to SSI exhibition. Besides that, by increasing the stop-band (equivalently to increase ξ_2) the tendency to exhibit SSI is

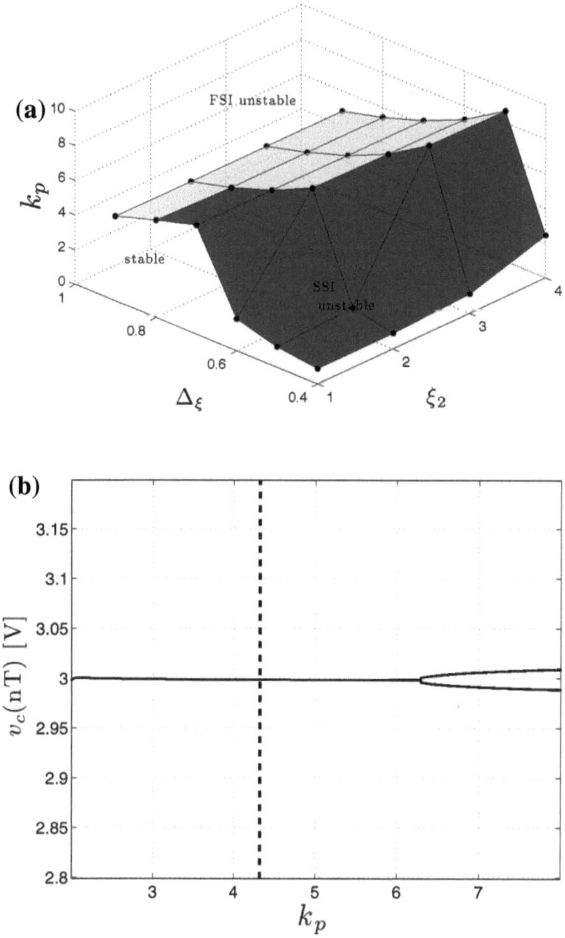

Fig. 4.12 **a** FSI boundary surface, obtained from the discrete-time model, in a VMC buck converter with a notch controller tuned at half of the switching frequency, as a function of attenuation Δ_ξ and ξ_2. FSI (*white*) and SSI (*black*) boundaries. **b** Bifurcation diagram as a function of the proportional gain k_p with parameters $\xi_1 = 1.2$, $\xi_2 = 2$ and $\Delta_\xi = 0.6$

reduced. It is worth mentioning that by increasing ξ_2, hence increasing the attenuation band, the results are close to those obtained from a simple PI compensator buck converter without chaos controller, the critical proportional gain of which, before FSI is exhibited, is $k'_{p,crit} = k_{p,crit}/\Delta_\xi$.

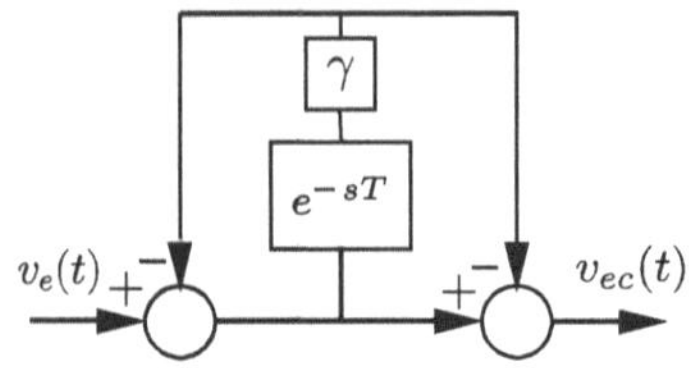

Fig. 4.13 Feed-forward repetitive controller structure

4.4 Repetitive Chaos Controllers

The previous controllers are based on attenuating the frequency content at half of the switching frequency to extend the stability margin in terms of FSI.

However, the use of a so-called repetitive controller, shown in Fig. 4.13, has been also used for chaos control purpose (Escobar et al. 2006; Corradini et al. 2008). The s-domain representation of such a repetitive controller as a function of parameter γ is:

$$G_{FS}(s) = \frac{e^{-sT} + \gamma}{e^{-sT} - \gamma} \tag{4.3}$$

The frequency response of a repetitive controller is shown in Fig. 4.14. Note that comparing with a simple TDFC, apart from attenuating the harmonic at half of the switching frequency, the controller is amplifying not only at the switching frequency harmonics but also at the DC component, hence losing the *non-invasive* nature that previous controllers had.

The bifurcation diagrams as a function of γ and the proportional gain k_p are shown in Fig. 4.15, which can be compared with the TDFC in Figs. 4.5 and 4.6 respectively, showing that the repetitive controller is less stable in terms of SSI but more stable in terms of FSI. Regarding the SSI observed results, they agree with the frequency domain interpretation, in which more phase lag is added by the controller. However, in terms of FSI, the improvement of the stability margin compared to TDFC can be only attributed to the amplification of switching frequency harmonic. These results are in agreement with what was observed in the frequency model derived in Sect. 3.8, in which the obtained stability condition (Eq. (3.99)) points out that the system is stable if the harmonic level at the switching frequency (proportional to voltage ripple) is higher than the level of the harmonic at half of the switching frequency.

4.5 Narrow Band Amplifier Chaos Controller

The previous sections, along with the frequency model derived in Sect. 3.8, suggest that FSI control can be accomplished not only by attenuating the harmonic at half of the switching frequency but also amplifying the switching frequency harmonic.

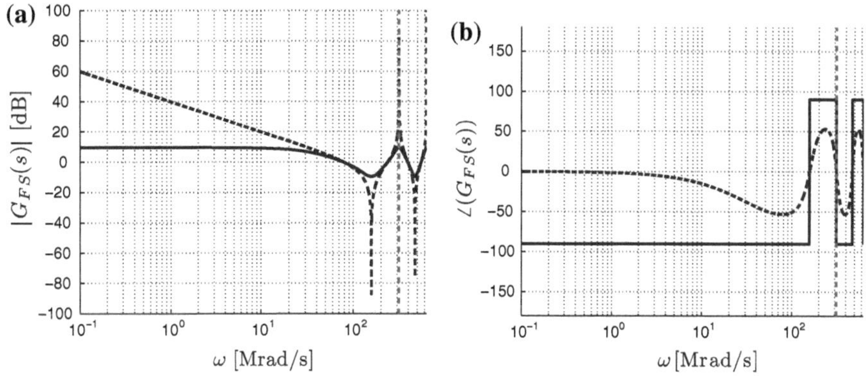

Fig. 4.14 Bode diagram representation (magnitude and phase) of the repetitive controller with $\gamma = 1$ (*solid*) and $\gamma = 0.5$ (*dash*)

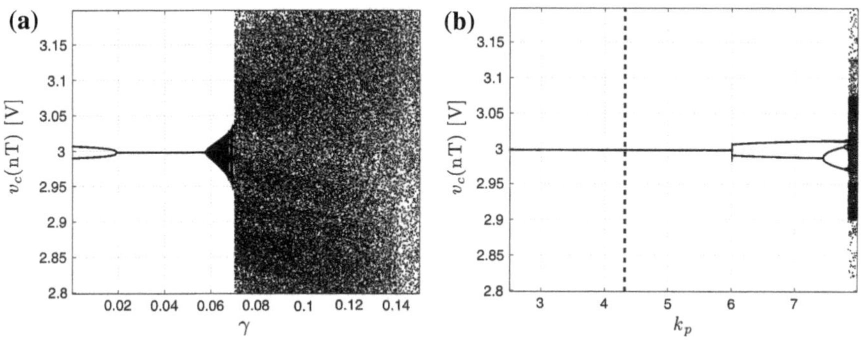

Fig. 4.15 Bifurcation diagram by sweeping **a** γ and **b** the proportional gain k_p in a VMC buck converter with a repetitive controller

Thus, this section is focused on providing the FSI control functionality from a feasible implementation of such harmonic amplification. This is carried out by using narrow-band amplifier (NBA) centered at the switching frequency. The complete block diagram is shown in Fig. 4.16. Note that in (Redl and Sun 2009), enhanced ripple regulators are proposed based on adding a single feedback path with an amplifier that improves the DC regulation while also "amplifies and shapes" the ripple voltage. The work points out that by means of such amplifier, based on a PID, the controller can improve the FSI. However, in that work it is not justified why FSI is eliminated (a clear cause of the effect of such amplifier upon the FSI is not provided) and the approach is not applied for PWM.

The transfer function of the FSI chaos controller $G_{FS}(s)$ can be described as in the case of a notch filter in Eq. (4.2) but with $\Delta_\xi > 1$ and the center frequency tuned to the switching frequency, $\omega_n = 2\pi f_s$. The magnitude and phase of such transfer function in the frequency domain is shown in Fig. 4.17. Apart from the amplification

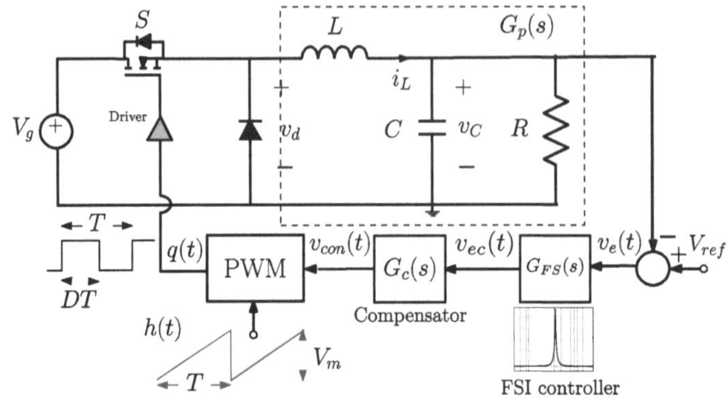

Fig. 4.16 **a** VMC buck converter with a NBA FSI controller

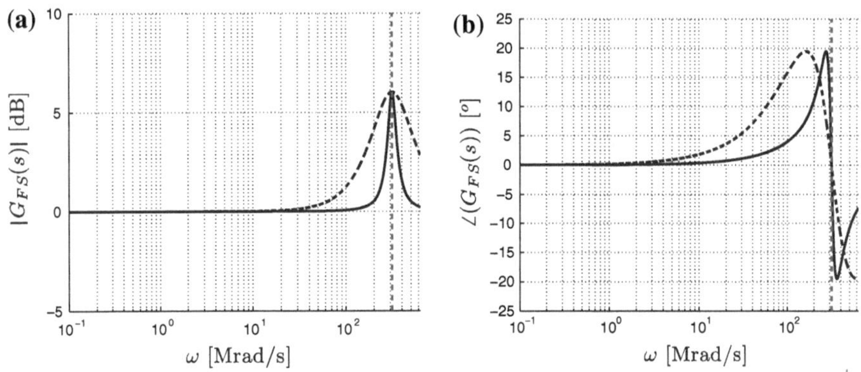

Fig. 4.17 Bode diagram representation (magnitude and phase) of a NBA tuned to the switching frequency (*vertical dashed-line*) with $\Delta_\xi = 2$ and $\xi_2 = 0.1$ (*solid*) and $\xi_2 = 0.5$ (*dash*)

at the switching frequency harmonic, a key additional advantage is the fact that it notably improves the SSI boundary by adding a phase lead before the switching frequency (note that no additional phase is added at the switching frequency).

Its effect upon FSI boundary is characterized in Fig. 4.18, which shows the stability boundary surface as a function of Δ_ξ and ξ_2. The surface has been obtained from the discrete-time model derived in Appendix A. On the one hand, the parameter ξ_2, which has a direct effect upon the width of the amplification band, has an important effect upon the stability boundary so that as it increases the FSI boundary is worsened. On the other hand, it is possible to observe that the higher the amplification is, the more stable the system is in terms of FSI.

The discussion and characterization of chaos controllers has been hitherto limited to the proportional gain k_p since it is one of the parameters that affects high-frequency magnitude but without modifying the phase. As it was mentioned at the beginning of this chapter, the indirect aim of a chaos controller, apart from obviously improving

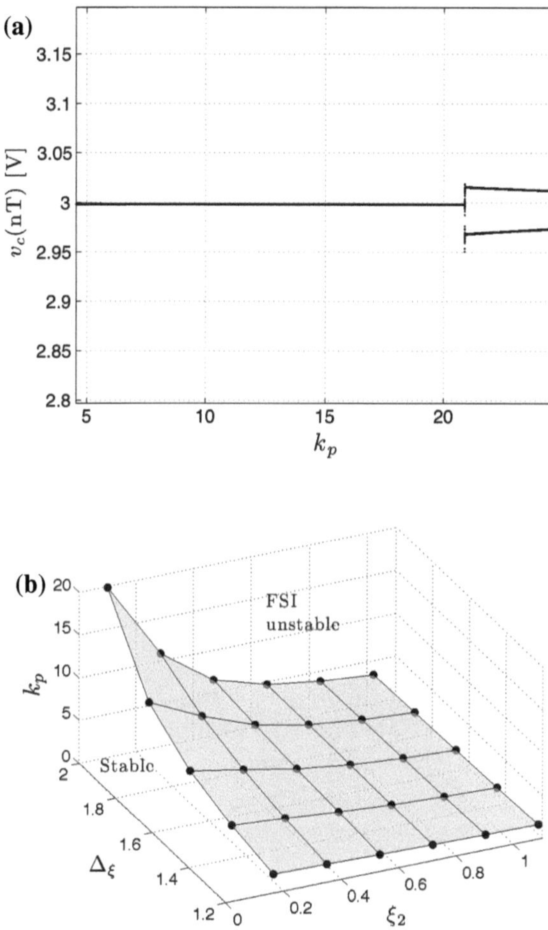

Fig. 4.18 **a** Bifurcation diagram by sweeping the proportional gain k_p of a VMC buck converter with a NBA controller ($\Delta_\xi = 2$ and $\xi_2 = 0.1$). **b** Stability surface, obtained from the discrete-time model as a function of the proportional gain k_p and parameters ξ_2, and Δ_ξ

the stability margin, is that such improvement allows to expand the design-space towards miniaturization (for instance reducing the inductance value, which is related to the circuit occupied volume), without losing stability.

Simulations in Fig. 4.19 show the stability boundary as a function of the inductance (related to area/volume occupancy) and switching frequency (related to efficiency). Note that both surfaces are similar and the stability boundary is clearly improved when $\xi_2 < 1$ and $\Delta_\xi > 0$ (a narrowband amplification) as compared to the case of not using such controller.

Finally, in the previous chapter, it has been shown that the FSI boundary strongly depends upon the duty cycle. Exploration of the effect of such parameter as a function

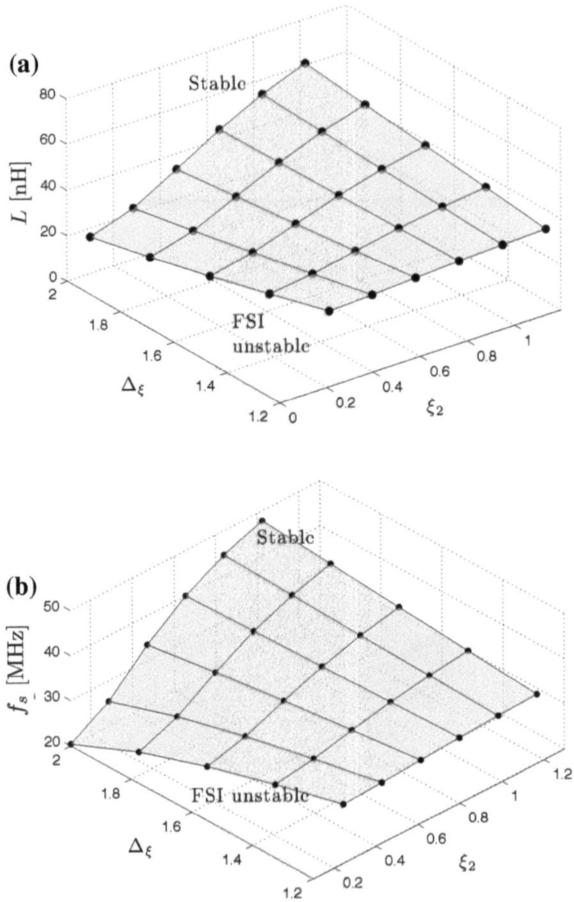

Fig. 4.19 FSI boundary surfaces, obtained from the discrete-time model, of a VMC buck converter with a NBA as a function of parameters Δ_ξ, ξ_2 and **a** the inductance L and **b** the switching frequency f_s

of the inductance has been carried out in Fig. 4.20 from the discrete-time model. The results show that the major benefits of such controller are shown for a duty cycle of 0.5 (note that the inductance can reach the boundary between continuous and discontinuous conduction mode without losing stability). The results are in agreement with the results in a conventional buck converter in the sense that for $D = 0.5$, the system is more stable.

Compared to the precedent chaos controllers, the NBA shows benefits in terms of overall stability, considering both FSI and SSI boundaries.

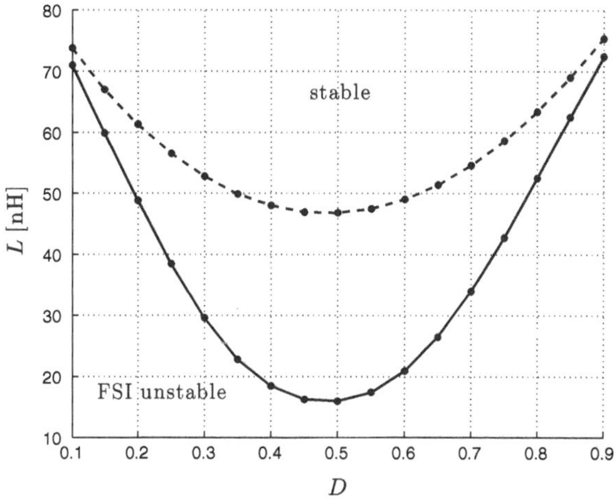

Fig. 4.20 FSI boundary curves, obtained from the discrete-time model, of a VMC buck converter with a NBA (*solid*) with $\xi_2 = 0.1$ and $\Delta_\xi = 2$ and without using the NBA controller (*dash*) as a function of the inductance L and the duty cycle D. $L_{\mathrm{DCM}} = 12.5\,\mathrm{nH}$

4.6 Towards a Low-Ripple High-Stability Regulation: Combining Chaos Controller with Output Ripple Reduction

The improvement in stability terms when a chaos controller is used implies a clear step forward towards circuit miniaturization without losing stability. However, this will imply an increase in the converter output ripple, which is not desirable in order to keep the DC-DC nature of the system. Therefore, the advantage of the NBA chaos controller could be invalidated because of the expansion of the design-space, namely area or switching frequency reduction could be limited by load specifications in terms of ripple.

The natural way for tackling this excess of ripple would be to add a ripple attenuation module to the converter output, but this is not trivial since this should be based on modifying the switching frequency harmonic, which is the same work principle of the NBA chaos controller, and hence it could cancel the advantage of such approach.

This section proposes a joint approach to obtain a high stability margin in terms of FSI, that allows reducing the converter reactive components, and at the same time reducing the output voltage ripple to accomplish low-ripple performance as required by DC-DC converters.

Two possible configurations with the aim of reducing the output ripple are shown in Fig. 4.21. The first one is based on adding an impedance zero at the switching frequency, hence attenuating the ripple. However this effect can cancel the amplification at the switching frequency carried out by the chaos controller. A controller, based on this approach, will be tackled in the next section.

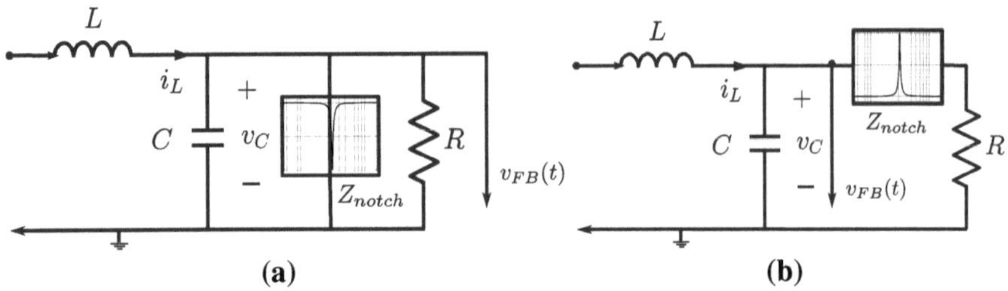

Fig. 4.21 Two possible configurations to reduce the output voltage ripple: **a** low impedance at the switching frequency from output to ground; **b** high impedance at the switching frequency from the feedback sensing point to load

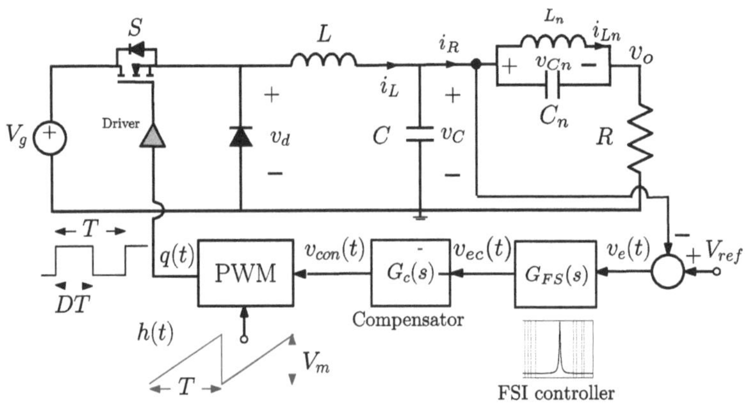

Fig. 4.22 VMC buck converter with a NBA chaos controller along with an output notch filter

Alternatively, the other approach is based on the elimination of output ripple by adding a high output impedance notch at the switching frequency. One possible implementation of this configuration is shown in Fig. 4.22. The approach does not modify the frequency response from the feedback standpoint and reduces the converter output ripple. The benefits of such output notch configuration in terms of area/ripple have been discussed in (Alarcon et al. 2004) as a solution to reduce the reactive component size.

The converter transfer function from the LC filter input v_d to the feedback voltage v_C and from the LC filter input v_d to the output v_o, namely $G_v(s)$ and $G_o(s)$ respectively, are:

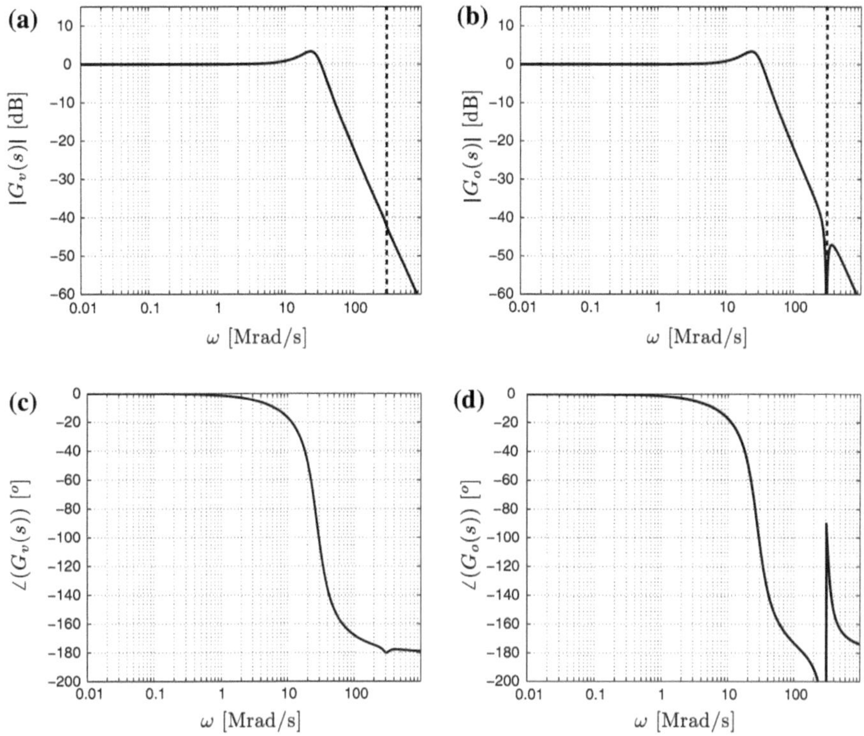

Fig. 4.23 Bode diagram representation (magnitude and phase) of a VMC buck converter with the output notch filter shown in Fig. 4.22. **a–c** Input-to-feedback transfer function $G_v(s)$ and **b–d** input-to-output transfer function $G_o(s)$

$$G_v(s) := \frac{v_C}{v_d} = \frac{\omega_0^2}{s^2 + \omega_{RC}\dfrac{R}{Z_o(s)}s + \omega_0^2} \tag{4.4}$$

$$Z_o(s) := \frac{v_C}{i_R} = R\frac{s^2 + \dfrac{s}{C_n R} + \omega_n^2}{s^2 + \omega_n^2} \tag{4.5}$$

$$G_o := \frac{v_o}{v_d} = G_v(s)\frac{R}{Z_o} \tag{4.6}$$

where $\omega_n = 1/\sqrt{L_n C_n} = 2\pi f_s$ (notch tuned at the switching frequency).

Both the input-to-feedback $G_v(s)$ and the input-to-output $G_o(s)$ transfer functions are shown in Fig. 4.23, in which it is shown that the output notch does not affect the feedback transfer function but adds a notch in the output transfer function hence reducing the amount of ripple delivered to the load. Additionally the NBA will be added as a chaos controller in the feedback loop.

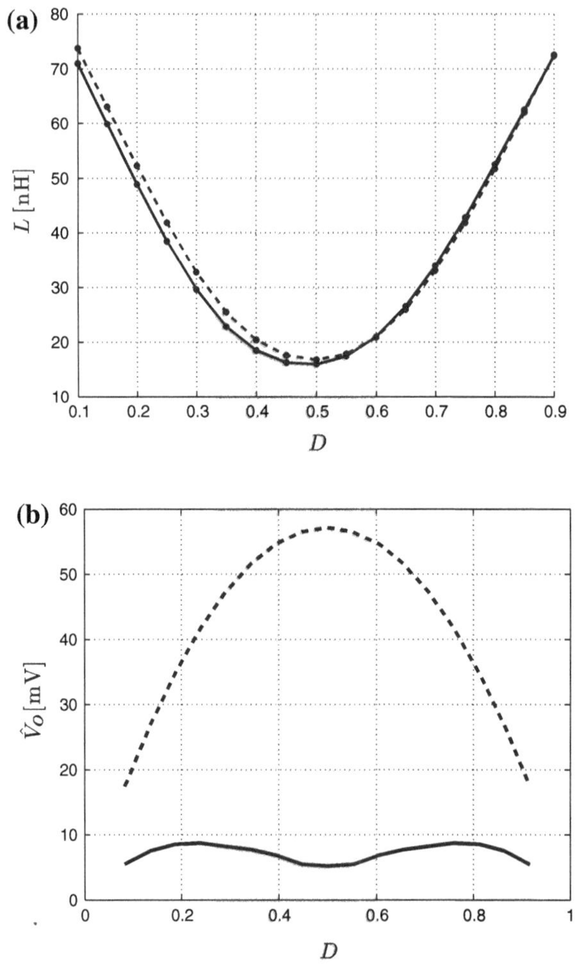

Fig. 4.24 **a** FSI boundary curves, obtained from the discrete-time model, of a VMC buck converter with a NBA controller and output notch filter (*solid*) and only with a NBA controller (*dash*) as a function of the inductance L and the duty cycle D and **b** the output voltage ripple $\hat{V}_o$ as a function of the duty cycle D. $L_n = 2\,\text{nH}$ $C_n = 5\,\text{nF}$, $\xi_2 = 0.1$ and $\Delta_\xi = 2$

In Fig. 4.24 results from using only the NBA (obtained in the last section) and this approach are compared. The results are obtained from the discrete-time model developed in Appendix A. Note that a similar stability boundary is obtained as a function of the duty cycle, but the output voltage ripple is largely reduced by canceling the harmonic at the switching frequency.

The main drawback of this proposed configuration is its additional required circuitry: the NBA plus the reactive components of the output notch. In addition, the notch inductor should drive all the output current hence requiring large area and the

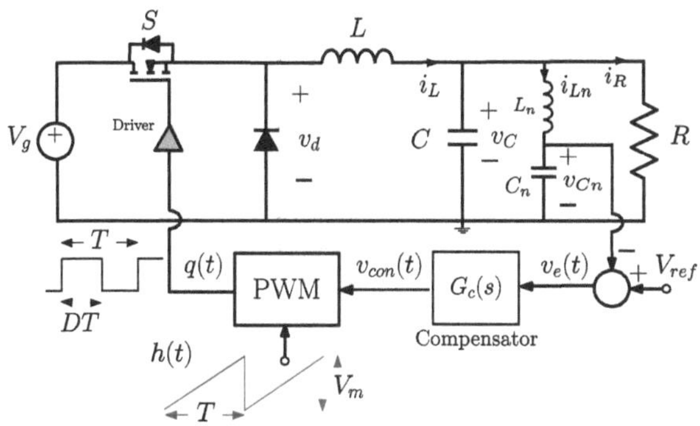

Fig. 4.25 VMC buck converter by sensing the output voltage ripple through an LC divider

circuit also requires accurately controlling the center frequency and the gain of the amplifier and the output notch since a mismatch between both center frequencies (such as due to the effect of temperature or parasitic elements) could limit benefits of the approach in terms of both output voltage ripple and fast-scale stability.

4.6.1 The LC Divider: Combining the Low Ripple and FSI Controller

An alternative approach to avoid FSI and to improve the output voltage ripple is explored in this section. The starting point is based on adding a low-impedance notch filter at the output of the converter, as it was shown in Fig. 4.21.

However, an output notch at the switching frequency directly connected to the same point as where the feedback voltage is sensed will reduce the voltage ripple at the output but also in the feedback loop, hence canceling the amplifying effect and limiting the stability benefits.

However, the approach proposed in this section, shown in Fig. 4.25 is based on filtering the output ripple by means of an LC divider, so that it consists of a notch filter function from the output voltage standpoint and, at the same time, changes the feedback voltage sensing point so that the harmonic at the switching frequency is amplified respect to the harmonic at half of the switching frequency, hence improving the FSI boundary.

In this scheme, it is not required any amplifier and only two reactive components are added to the system. Furthermore, the output inductor of the notch impedance does not drive all the output current thus facilitating its integration. The system parameter values used in this section are $V_g = 6\,\text{V}$, $V_{ref} = 1.2\,\text{V}$, $V_m = 1\,\text{V}$, $f_s = 50\,\text{MHz}$, $\text{R} = 2.5\,\Omega$, $k_p = 3$, $L = 66\,\text{nH}$, $C = 20\,\text{nF}$, $\omega_{z1} = 1\,\text{Mrad/s}$ and $L_n = 2\,\text{nH}$ and $C_n = 5\,\text{nF}$.

The transfer function of both paths, from the LC filter input v_d to the output voltage v_o and from the LC filter input v_d to the feedback voltage v_{Cn}, $G_o(s)$ and $G_c(s)$ respectively, are given by:

$$G_o(s) := \frac{v_o}{v_d} = \frac{\omega_0^2}{s^2 + \omega_{RC}\omega_0 \frac{R}{Z_o(s)} s + \omega_0^2} \tag{4.7}$$

$$Z_o(s) := \frac{v_C}{i_{Ln} + i_R} = R \frac{s^2 + \omega_n^2}{s^2 + \frac{R}{L_n} s + \omega_n^2} \tag{4.8}$$

$$G_v(s) := \frac{v_{Cn}}{v_d} = G_o(s) \frac{\omega_n^2}{s^2 + \omega_n^2} \tag{4.9}$$

where $\omega_n = 1/\sqrt{L_n C_n} = 2\pi f_s$ (notch tuned to the switching frequency).

The frequency domain representation of these transfer functions are depicted in Fig. 4.26.

The output transfer function $G_o(s)$ includes a notch at the switching frequency, hence reducing the output ripple, whereas in the feedback path $G_v(s)$ the harmonic at the switching frequency is amplified relative to the harmonic at half of the switching frequency.

The benefits of the controller in terms of FSI exhibition are depicted in Fig. 4.27, in which it can be observed that the controller clearly improves the stability boundary as a function of k_p. Note that due to the effect of the proposed controller, another type of instability is exhibited instead of the classical period-doubling behavior.

As it was demonstrated during this chapter, the improvement in FSI boundaries allows varying some parameters without losing stability, such as the inductance value or the switching frequency, which are of especial interest for miniaturization.

The FSI boundary surface, obtained from the discrete-time model derived in Appendix A, as a function of the inductance L and the duty cycle D is shown in Fig. 4.28 and it is compared to the one obtained from a conventional buck converter (without any FSI controller). Note that the LC divider approach results in better stability boundaries, allowing to reduce the inductance value without losing stability and being its minimum value limited by the DCM border in a wide range of duty cycles. Similar design space as a function of the switching frequency demonstrates similar benefits of such controller. Note that modifying the switching frequency implies also modifying the an LC divider (notch) resonant frequency.

Furthermore, Fig. 4.28 also presents a comparison between the output ripple obtained from LC divider buck converter and from a conventional buck converter showing clear benefits that will facilitate the reduction of the reactive components parameter values or the switching frequency.

One of the main drawbacks of such compact structure is the dependence that both metrics, ripple and stability, can have upon the reactive components tolerance and quality factor.

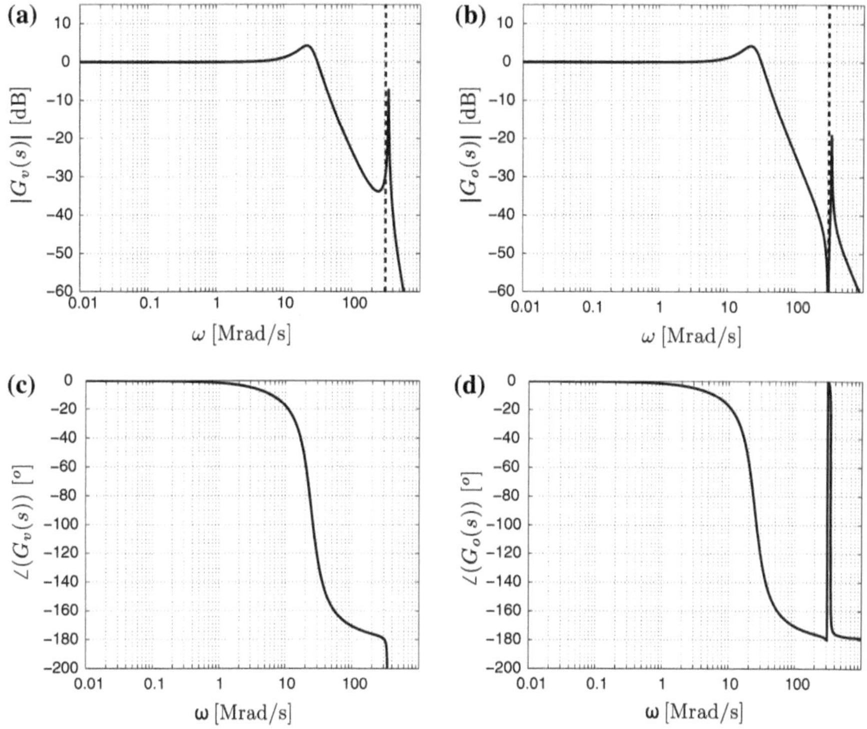

Fig. 4.26 Bode diagram representation (magnitude and phase) of a VMC buck converter with the LC divider in Fig. 4.25. **a–c** Input-to-feedback transfer function $G_v(s)$ and **b–d** input-to-output transfer function $G_o(s)$

The effect of the tolerance on the ripple and the stability is characterized in Fig. 4.29 in which parameters L_n and C_n are modified with a tolerance of 5 %. Regarding the stability, it is possible to observe that as the feedback resonant peak is closer to the switching frequency, the system is more stable in terms of FSI (this corresponds to an increase of parameters values C_n and L_n), but if this accurate resonant frequency can not be reached, due to frequency drifts, temperature among other factors, the stability surface is very flat, almost not affecting such boundary. Regarding the effect of reactive components tolerance upon the output voltage ripple, it can be considerable, as it is shown in Fig. 4.29. At nominal values there is a minimum of the output ripple and a variation on both sides leads to its increase. Despite this, note that the output voltage ripple is still considerably low compared to the output voltage ripple this converter would have without an LC divider structure.

Furthermore, the effect of the reactive components quality factor (Q_{Cn}, Q_{Ln}) is also characterized in Fig. 4.30. The quality factor of both reactive components are defined as:

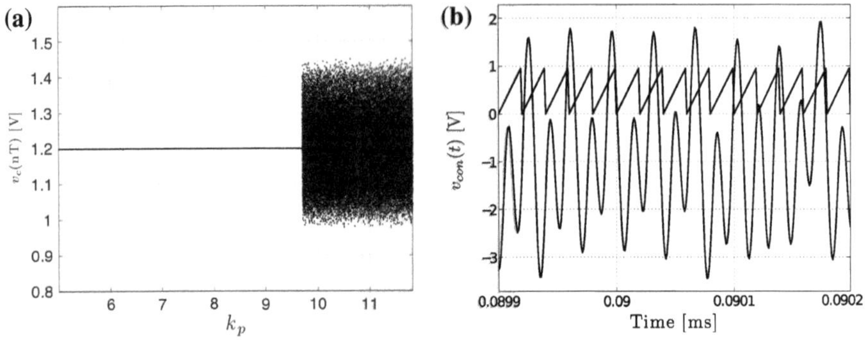

Fig. 4.27 **a** Bifurcation diagram by sweeping the proportional gain k_p of a VMC buck with an LC divider. **b** Control signal at the modulator input waveform v_{con} after bifurcation takes place. $k_{p,crit} = 3.22$ (without LC divider)

$$Q_{Cn} = \frac{1}{\omega_s R_{Cn} C_n} \tag{4.10}$$

$$Q_{Ln} = \frac{\omega_s L_n}{R_{Ln}} \tag{4.11}$$

Note that the quality factors of both reactive components have a similar effect upon the stability margin.

4.6.2 Application to CMC

The application of the LC divider chaos controller to the CMC case is considered in this section, instead of using an external ramp. The circuit diagram is illustrated in Fig. 4.31.

Figure 4.32 shows an example in which FSI is exhibited in a CMC buck converter. As it can be observed, the LC divider, apart from attenuating the output voltage ripple, maintains the system stable. The parameters are the same as the previous ones but with $V_{ref} = 1.5\,\text{V}$ and current gain $k_i = 2$.

As it was demonstrated in Sect. 3.7, the FSI in CMC has a strong dependence upon the duty cycle. Therefore in Fig. 4.33, it is explored the effect of such parameter upon the stability boundary, obtained from the discrete-time model developed in Appendix A.6, as a function of the voltage-feedback proportional gain k_p.

It can be observed that the stable behaviour is reached by increasing the proportional gain k_p. Then, for duty cycles below 0.5, the effect of the LC divider cancels the possibility of FSI exhibition, hence not requiring any feedback gain. When the duty cycle is higher than 0.5, the control of such instabilities requires increasing the proportional gain to reach a stable behavior. This trend can be observed in Fig. 4.33, in which it is shown the by sweeping the for $V_{ref} = 4\,\text{V}$.

Fig. 4.28 FSI boundary curves, obtained from the discrete-time model, of a VMC buck converter with an LC divider (*solid*) and without the LC divider (*dash*) as a function of the duty cycle D and **a** inductance L and **b** switching frequency f_s **c** the output ripple $\hat{V}_C$ as a function of the duty cycle D

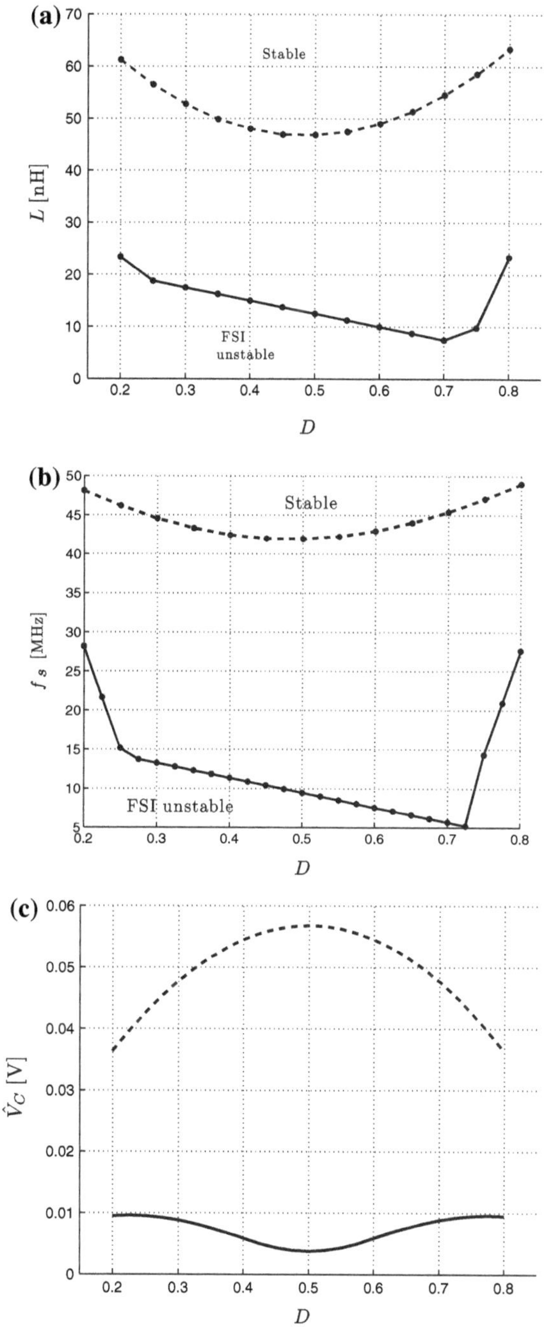

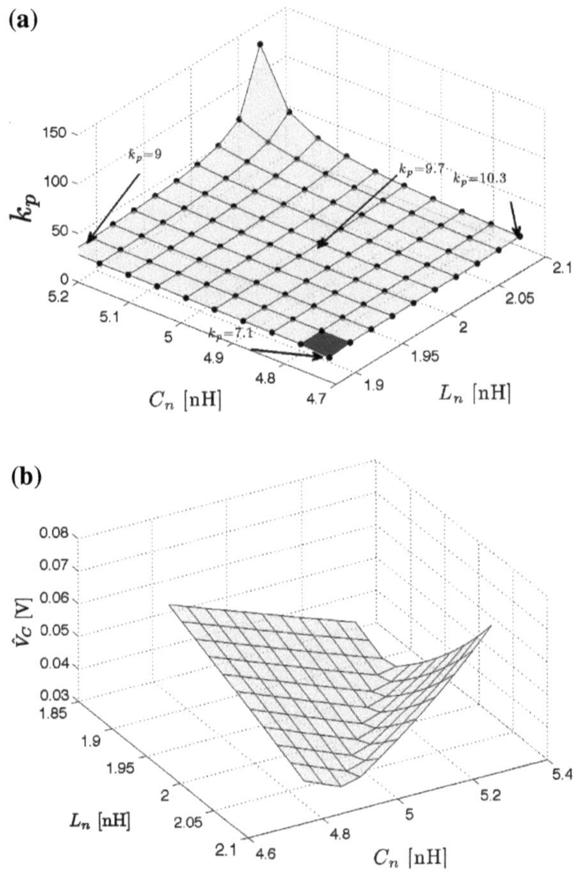

Fig. 4.29 **a** FSI boundary surface, obtained from the discrete-time model, and **b** the output voltage ripple $\hat{V}_C$ as a function of the tolerance of the reactive component parameters. Nominal values $C_n = 5\,\text{nF}$ $L_n = 2\,\text{nH}$. $k_{p,crit} = 3.22$ (without controller)

Note that the fact that increasing the proportional gain leads the converter to be more stable is opposite to what was observed in Sect. 3.7, in which the increase of the ripple led to FSI. This discrepancy can be understood by analyzing the equation used to derive the closed-form stability condition in CMC, given in Eq. 3.58. Such stability boundary equation (without considering the ramp slope) can be written as:

$$k_v 4 f_s \hat{V}_c - \frac{k_v V_g}{4LCf_s} + k_i \dot{i}_L(DT^-) - \frac{k_i V_g}{2L} = 0 \Rightarrow \Delta_v - \Delta_i = 0 \qquad (4.12)$$

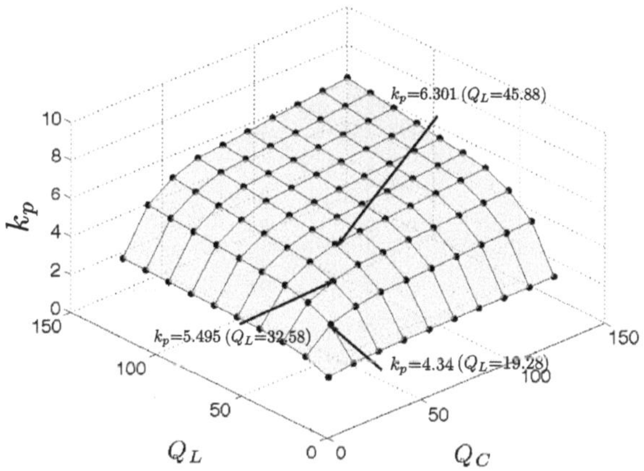

Fig. 4.30 FSI boundary surface, obtained from the discrete-time model, as a function of the quality factor of reactive component parameters. Nominal values $C_n = 5\,\text{nF}$, $L_n = 2\,\text{nH}$. $k_{p,crit} = 3.22$ (without LC divider controller)

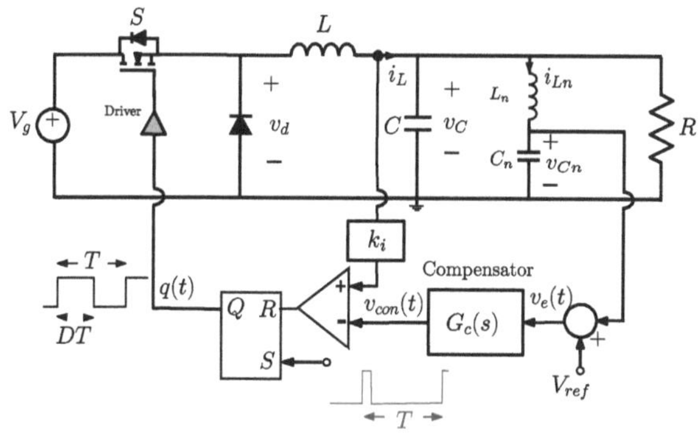

Fig. 4.31 CMC buck converter by sensing the output voltage ripple through an LC divider

where

$$\Delta_v = \frac{k_v V_g}{2LCf_s}\left(D\overline{D} - \frac{1}{2}\right) \tag{4.13}$$

$$\Delta_i = \frac{k_i V_g}{L}\left(\frac{1}{2} - D\right) \tag{4.14}$$

The stability condition is $\Delta_v - \Delta_i > 0$. Note that for $D < 1/2$, $\Delta_v < 0$ and $\Delta_i > 0$, the current loop is compensating the voltage loop and improving the stability. On the other hand, for $D > 1/2$ both parameters are negative and hence the system becomes unconditionally unstable in terms of FSI and thus it requires an external ramp to obtain a stable behavior. However, if the LC divider is added (or an equivalent NBA), which amplifies the feedback voltage ripple, Δ_v would be higher than 0, and then for $D < 0.5$ both terms will be positive, so that the system becomes

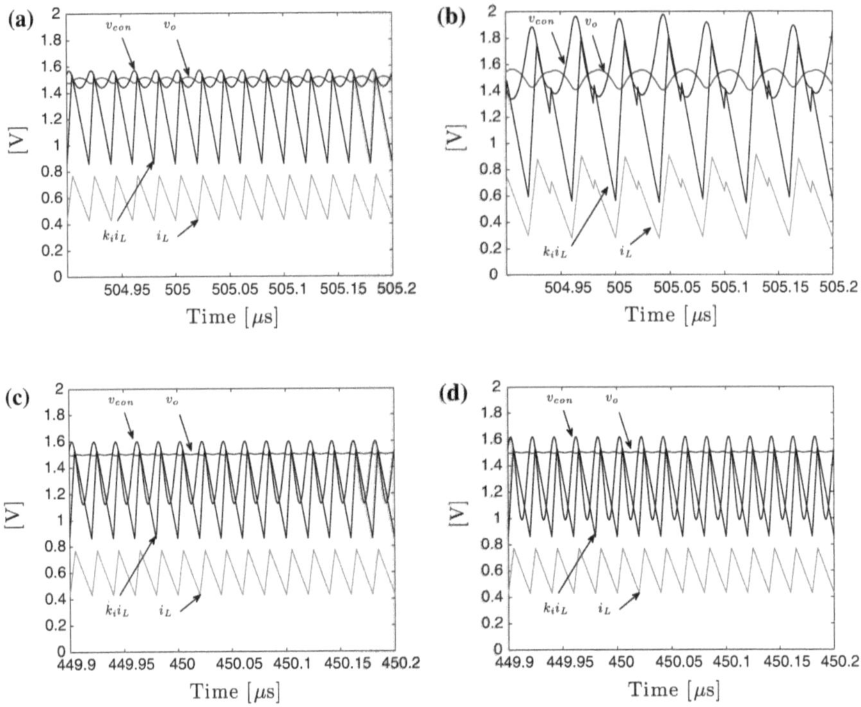

Fig. 4.32 Time-domain waveforms of the state variables in a CMC buck converter without (*top*) and with (*bottom*) LC divider chaos controller and $k_p = 3$ (*left*) and $k_p = 4$ (*right*)

unconditionally stable independently of the system parameters, and for $D > 0.5$ the system becomes stable provided that the gain added by the control makes Δ_v to be high enough to compensate the negative value of Δ_i, hence requiring to increase the feedback voltage ripple.

4.7 Stability Margins and Power Metrics Comparative Between Controllers

Up to now, different FSI controller approaches have been explored individually in terms of stability. This section combines all approaches and compares them not only in terms of stability, but also in terms of power metrics, namely area, efficiency, ripple, and dynamic regulation. The parameters used in this section are the same as in the previous sections: $V_g = 6$, $V_{ref} = 1.2\,\text{V}$, $C = 20\,\text{nF}$, $L = 66\,\text{nH}$, $R = 2.5\,\Omega$, $k_p = 3, \omega_{z1} = 1\,\text{Mrad/s}$, $f_s = 50\,\text{MHz}$ and $V_m = 1\,\text{V}$.

The stability boundaries comparison, obtained by numerical simulations along with the discrete-time model, of TDFC, ETDFC, notch-based controller, NBA and LC divider is shown in Fig. 4.34. This comparison has been carried out by sweeping

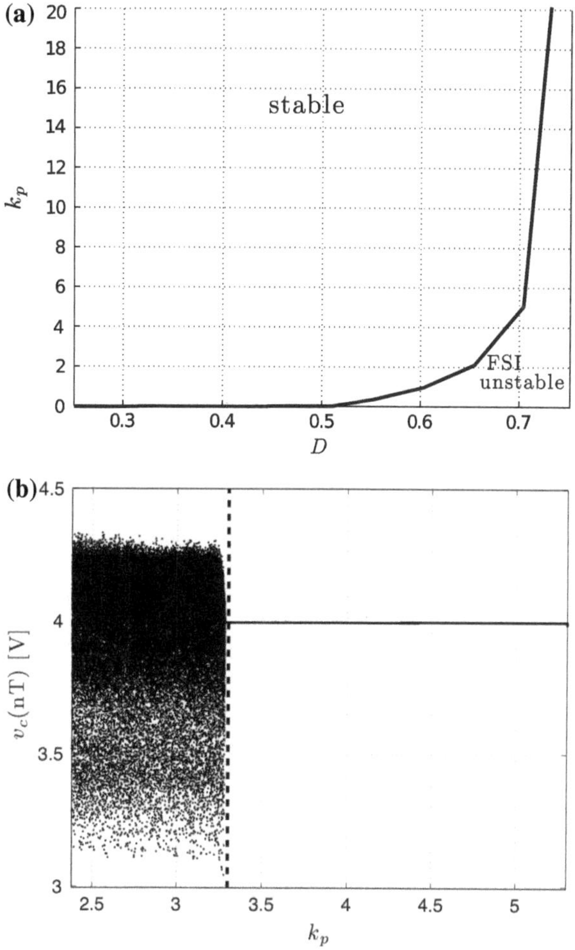

Fig. 4.33 **a** Stability boundary obtained from the discrete-time model by sweeping the duty cycle D. **b** Bifurcation diagram by sweeping the proportional gain k_p in a CMC buck converter with $V_{ref} = 4\,\mathrm{V}$

the PI compensator parameters, namely the proportional gain k_p and the zero ω_{z1}, due to their capability to explore a wide design-space range including both FSI and SSI boundaries.

The exploration shows that delay-time-based controllers have a limited benefits, because of their negative effect upon SSI boundary, hence reducing their advantage in overall stability terms. It is interesting to observe that the notch-based approach, which has been derived from such time-delay-based controllers, by properly adjusting the damping factors (ξ_1 and ξ_2), can result in improved behavior.

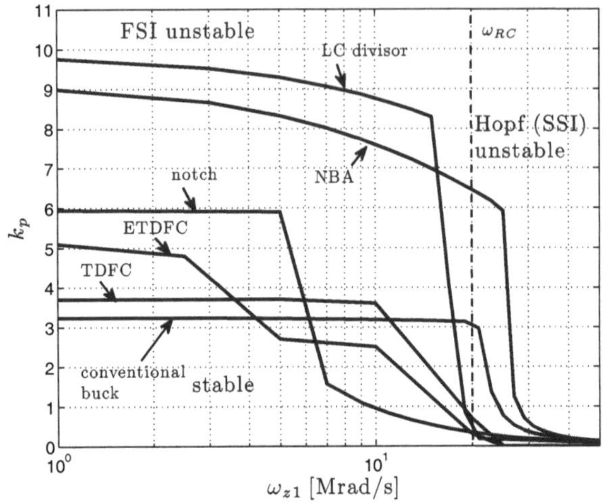

Fig. 4.34 Stability boundaries, obtained from the respective discrete-time model and numerical simulations, of a VMC buck converter as a function of the PI compensator parameters k_p and ω_{z1} by using different FSI controllers, adding a TDFC with $\gamma = 0.05$; a ETDFC with $\gamma = 0.2$ and $\beta = 0.5$; a notch with $\xi_1 = 1.2$ and $\xi_2 = 2$; a NBA with $\xi_1 = 0.1$ and $\xi_2 = 0.4$, and an LC divider with $L_n = 2\,\text{nH}$ and $C_n = 5\,\text{nF}$

The alternative approaches based on amplifying the switching frequency harmonic show a better behavior in terms of overall stability. The simple NBA is the only controller that concurrently improves both stability boundaries and the LC divider, reaches a considerable improvement in terms of FSI boundary, although the SSI boundary is slightly worsened.

FSI controller can not be only observed as "controllers" just affecting the stability but also as key low-ripple low-area enabling controllers so that they can facilitate the integration of switching power converters by means of ensuring stability for low-area and low switching frequency scenarios.

A controller with improved FSI behaviour is not only affecting the stability region, but it indeed expands the parameter space of feasible designs, thereby allowing to trade off stability versus ripple versus occupied area, so that these controllers can be interpreted as a means of facilitating the miniaturization and integration of switching power converters by ensuring stability for low area occupancy and low switching frequency.

However, a design shift towards miniaturization cannot be considered feasible if such occupied area or switching frequency reductions do not go along with a reduction of output voltage ripple. Therefore, a 3-metrics map is carried out in Fig. 4.35 to compare the design-space, obtained as a function of the inductance value (Area), the switching frequency (related to efficiency through the dominant switching losses) and the output voltage ripple ($\hat{V}_C$) for the application of the different alternative controllers. These surfaces have been obtained by fixing the switching frequency and obtaining the minimum value of inductance before transiting to DCM or losing

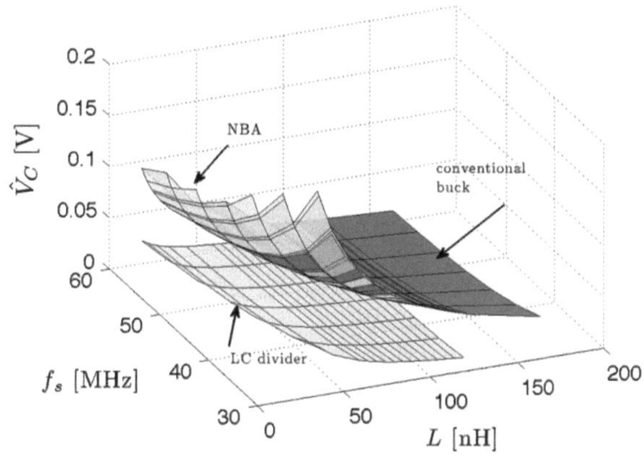

Fig. 4.35 Power metrics design-space surface as a function of switching frequency (efficiency), inductance (area) and ripple for different FSI controllers: a conventional buck converter; an NBA with $\xi_1 = 0.1$ and $\xi_2 = 0.4$, and an LC divider with $L_n = 2\,\text{nH}$ and $C_n = 5\,\text{nF}$

stability. Comparing the NBA and the conventional cases, it is worth observing that the NBA controller design-space surface is already extended, which indeed allows using lower inductance values without losing stability, but the ripple is equivalent in both controllers. However, observing the outcomes of the LC divider controller case, it can be stated that it clearly improves the overall performance as compared to the other counterpart controllers by extending the surface in a similar way as in the NBA, but at the same time reducing the output voltage ripple magnitude.

Beyond metrics such as steady-state stability or output voltage ripple, in regulation applications, it is very important to evaluate the dynamic response of the converter in front of a load change of demand. The simulation in Fig. 4.36 tackles the characterization of the controller performance from a dynamics standpoint. The interest is to qualitatively evaluate the response of the converter to output current step.

Figure 4.36 unveils that chaos controllers are not only of interest because of their fast-scale stability boundary improvement, but also since they can reduce the settling time when a load step occurs. As it can be observed, the NBA and LC divider controllers have a better transient response than the conventional buck regulator with the same parameter values, while the notch-based controller gives a similar performance.

4.8 Concluding Remarks

This chapter has addressed the synthesis of FSI controllers from an engineering design-oriented standpoint together with the aim of improving their interpretation so as to be able to enhance their performance and facilitate their implementation. First, the previously reported delay-based controllers have been revisited and, by characterizing that they are based on comb filtering tuned to half of the switching frequency and its multiple harmonics, it has been shown their limitations regarding overall stability, since controlling FSI is detrimental to the SSI boundary, as well as the

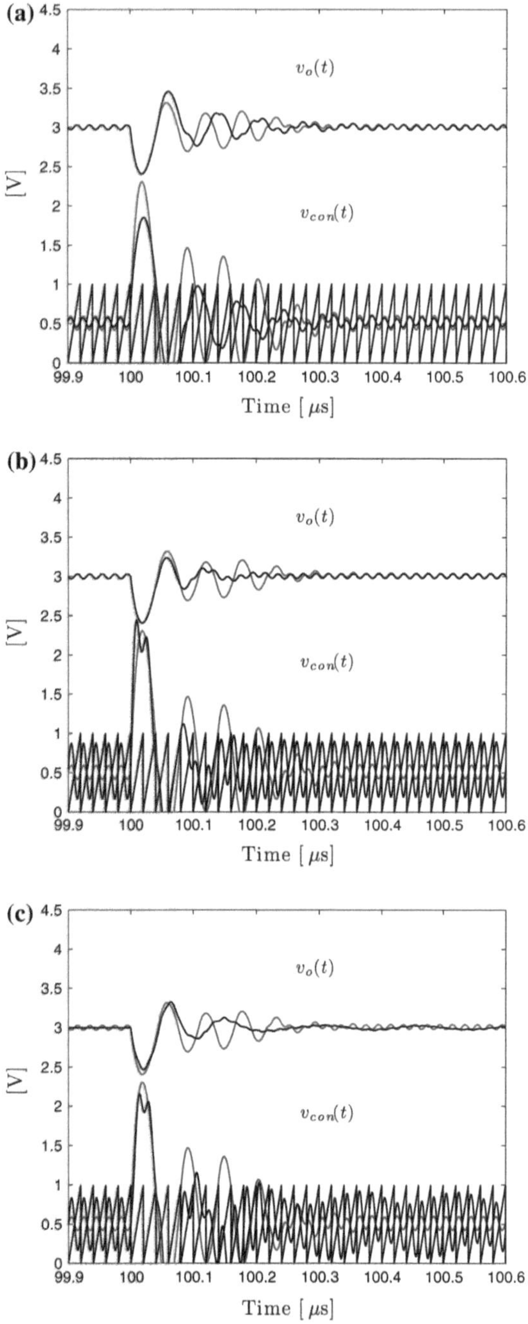

Fig. 4.36 Transient response of the VMC buck converter in front of an output current step from $R^- = 2.5$ to $R^+ = 1.25\,\Omega$ at $t_{0=100}\,\mu$s for a conventional buck (*red*) and different FSI controllers: **a** notch-based controller with $\xi_1 = 1.2$ and $\xi_2 = 2$; **b** NBA with $\xi_1 = 0.1$ and $\xi_2 = 0.4$ and **c** LC divider with $L_n = 2\,$nH and $C_n = 5\,$nF

difficulties in implementation. An alternative proposal has been proposed by simplifying them to a simple notch filter tuned at half of the switching frequency, for which even improved stability performance is obtained as compared to time-delay-based controllers. Subsequently, a complementary approach has been proposed inspired in repetitive controllers, which has shown that by also amplifying the switching frequency harmonic, the FSI can be better controlled. An implementation-aware controller has been proposed based on an NBA, which not only improves FSI but also SSI boundaries. The controller allows notably reducing different parameters such as the inductance (related to system area) or switching frequency (related to system efficiency) whilst guaranteeing stability. From the knowledge of such FSI controller, the study has been extended by taking into account other power processing performance metrics such as the output voltage ripple, with the ultimate aim of facilitating converter miniaturization. This chapter has proposed two approaches to improve overall stability at the same time as reducing the output voltage ripple. While the first one requires two independent modules to address the combined improvement in terms of both stability and ripple, the second one is based on a compact LC divider circuit structure which achieves both objectives. The LC divider shows improved performance regarding both an enhancement of the stability region and a reduction in the output voltage ripple, hence resulting in an expansion of the region of potential designs which results in improved performance in terms of miniaturization, namely, low area, high-efficiency and low-ripple.

Chapter 5
Extension to Alternative Topologies and Functionalities Aiming Power Management Integrated Circuits

The previous chapters have focused on addressing, respectively, the characterization, prediction and control of instabilities in a basic buck converter due both to its widespread use as well as its interest as a core topology for miniaturization. This chapter further explores and extends these approaches to more advanced topologies and applications, as encountered in an integrated power management context.

The first section addresses the characterization and prediction of FSI in a multilevel multiphase converter, a natural candidate for further miniaturization due to its inherent improvement of the ripple and switching frequency tradeoff. The study unveils that this topology can enhance the overall performance as compared to a conventional buck converter in terms of fast-scale stability, although these benefits can be limited if the floating capacitor is reduced.

Subsequently, the chapter addresses the characterization of fast-scale instabilities in a buck-based switching converter in a wideband tracking application, in particular focusing on the extended parameters in the design space, namely the modulating frequency. Furthermore, it provides a discrete-time map model for the tracking case, which allows reducing the computation time to characterize instabilities in the switching amplifier. Finally, it compares the results of the wideband switching amplifier in terms of stability boundaries with the regulation case, demonstrating that the stability boundary in regulation constitutes a conservative limit to ensure stability in tracking applications.

5.1 The Three-Level Buck-Based Converter: Characterization, Modeling and Prediction of Instabilities

The floating-capacitor multilevel converter (Meynard et al. 2002), also called multicell, is a good candidate both for miniaturization/on-chip integration (Villar and Alarcon 2008) and for wideband tracking applications (Yousefzadeh et al. 2005) due to its inherent enhanced performance within the design space composed by

E. Rodríguez Vilamitjana et al., *Chaos in Switching Converters for Power Management*,
DOI: 10.1007/978-1-4614-2128-3_5, © Springer Science+Business Media New York 2013

the ripple, the reactive component size, the switching frequency and the bandwidth (Yousefzadeh et al. 2005). However, common to any other switching power converter, the multilevel converter might exhibit various instabilities, in particular FSI, depending on the values of the design-space parameters.

The three-level buck converter, shown in Fig. 5.1, is based on four switches and an additional floating capacitor C_x. The figure also shows the time-domain waveforms of the LC-filter input v_x and the floating capacitor voltage v_{cx}. It is worth identifying the two kinds of periods that this converter has: on the one hand the system period T_{cycle}, which is the external ramp period, and the interval of time in which the system dynamic is repeated. On the other hand, the switching period $T = 1/f_s$, which is the interval of time within which commutation occurs, in which the control signal v_{con} hits any of both ramp waveforms, hence changing from a energy supplied to free-running configuration. As it can be observed in Fig. 5.1, $T_{cycle} = 2\,\mathrm{T}$.

The parameters values of this case study are: $V_g = 6$, $V_{ref} = 3\,\mathrm{V}$, $V_m = 1\,\mathrm{V}$, $f_s = 50\,\mathrm{MHz}$, $R = 2.5\,\Omega$, $L = 66\,\mathrm{nH}$, $C = 20\,n\mathrm{F}$, $C_x = 200\,\mathrm{nF}$ and $\omega_{z1} = 0.1\,\mathrm{Mrad/s}$.

The three-level converter reduces the output ripple (for the same L, C, f_s), compared to an elementary buck converter, by reducing the input voltage v_x that is applied to the LC filter.

The output voltage control is carried out by means of the voltage feedback loop, but to ensure that the converter works properly, it is also required that the average floating capacitor voltage remains at half of the input voltage, which can not be directly guaranteed upfront by means of such output voltage feedback. The natural balancing of such capacitor voltage has been tackled in previous works (Wilkinson et al. 2006) by using a frequency domain standpoint. The analysis of such natural balancing of the floating voltage in a three-level buck-based converter can be addressed by distinguishing between two cases: for the output voltage being below one half of the battery voltage and above such limit.

In the first case ($V_o < V_g/2$), the voltage balance does not require any external control to compensate such capacitor voltage since, inherently, due to the multilevel operation, such voltage is regulated.

This is illustrated in Fig. 5.2, in which it is shown the control voltage in a open-loop configuration (applying constant duty cycles) by fixing the floating capacitor voltage slightly over one half of the battery $v_{cx} > V_g/2$. In this case, it can be observed that the resulting control waveform looks like a period-doubling behavior, but it is not related to a system instability. This behavior results in a decrease of the duty cycle D_1 of the PWM waveform $q_1(t)$ and an increase the value of the duty cycle D_2 of the second PWM waveform $q_2(t)$. This fact, indeed, leads to regulate the floating capacitor voltage since decreasing D_1 and increasing D_2 entails reducing such voltage.

An equivalent compensation occurs when the floating capacitor voltage is fixed slightly below one half of the battery, hence naturally regulating it. However, for output voltages over one half of the battery, this inherent regulation of the floating capacitor does not occur, hence requiring to take into account the floating capacitor

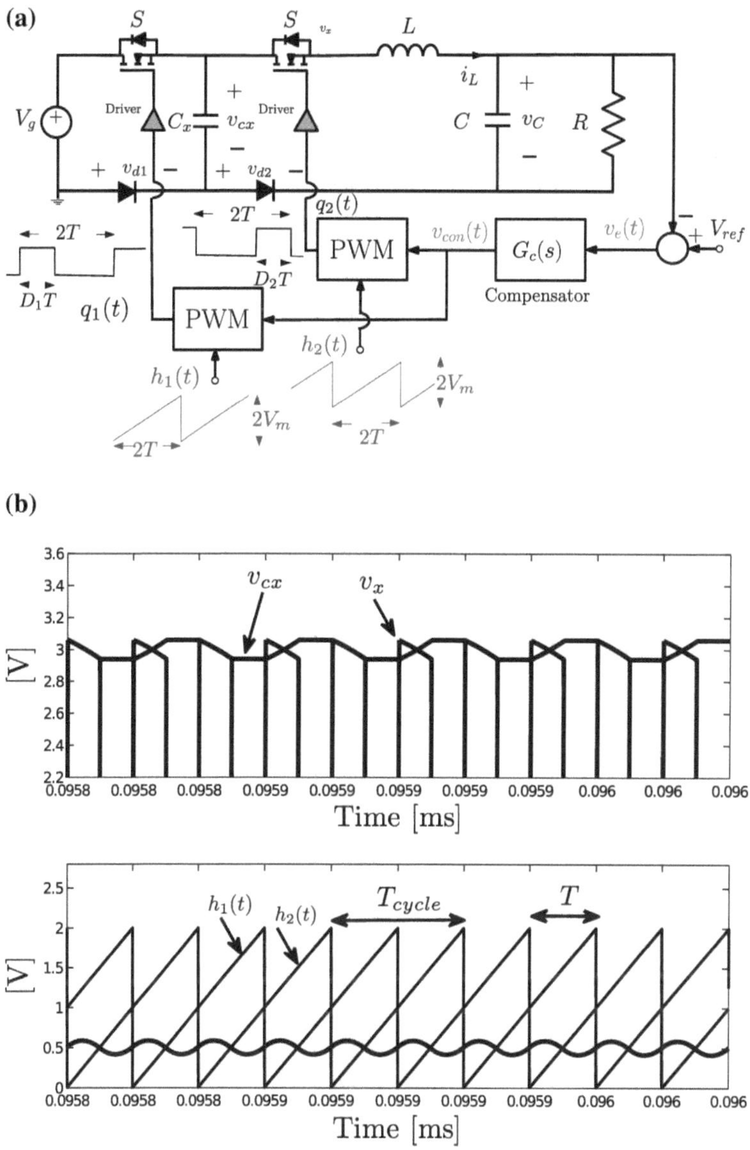

Fig. 5.1 **a** VMC three-level buck-based converter with controller $G_c(s)$ and PWM control. **b** (*top*) Time-domain waveform of floating capacitor voltage (v_{cx}) and LC-filter input signal (v_x) and (*bottom*) PWM modulator input and ramp

voltage in the control loop. The following section will focus on the study of the instabilities in the naturally balanced multilevel converter.

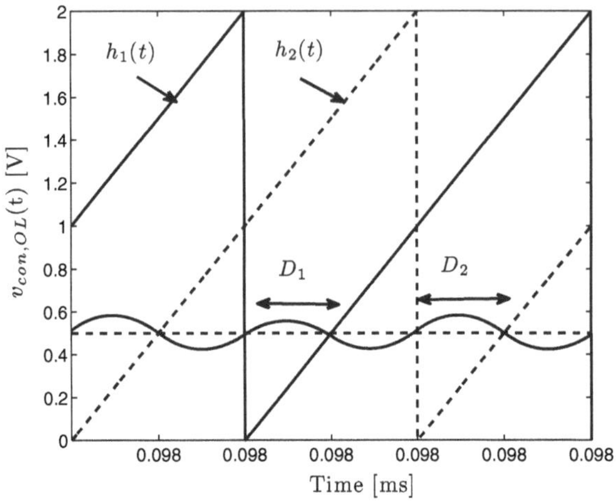

Fig. 5.2 Multilevel converter driven in open-loop with $V_{con} = 0.5$ being a DC external source voltage applied to the modulator input, with unbalanced floating capacitor ($v_{cx} > V_g/2$) where $v_{con,OL} = V_{con} + k_p(V_{ref} - v_C)$ would be the control voltage at the modulator input

5.1.1 Characterizing Instabilities in a Three-Level Buck Converter

In Fig. 5.3 the bifurcation diagram of the three-level converter by sweeping k_p is shown, obtained by sampling at the switching frequency the state variables (output capacitor and floating capacitor voltages) in steady-state regime.

By contrasting the behavior exhibited in Fig. 5.3 with that of the elemental buck converter bifurcation diagram shown along the book, the simulation shows the absence of classical period-doubling behavior. Such instability is characterized in Fig. 5.4, unveiling that the unstable behavior consists of SSI, as it can be observed in both time and spectral domains. Note that additional period-doubling-like behavior that coexists with the SSI can be due to the inherent behavior that the converter has when the floating capacitor voltage is not equal to $V_g/2$, previously shown in Fig. 5.2. Observing the associated spectra, it can be observed that while there is a harmonic spike at the switching frequency, additional harmonics components appears around half of the switching frequency.

5.1.2 Low Capacitance Value of the Floating Capacitor

Due to the demand for miniaturization, which implies the use of low values of the reactive components, it is of especial interest to characterize the effect of low values of the floating capacitance upon the stability.

Figure 5.5 shows the bifurcation diagram for $C_x = 20\,\text{nF}$ along with a characterization of the maximum proportional gain k_p in order to ensure the stable period-one behavior as a function of the floating capacitance C_x. This numerical simulation-

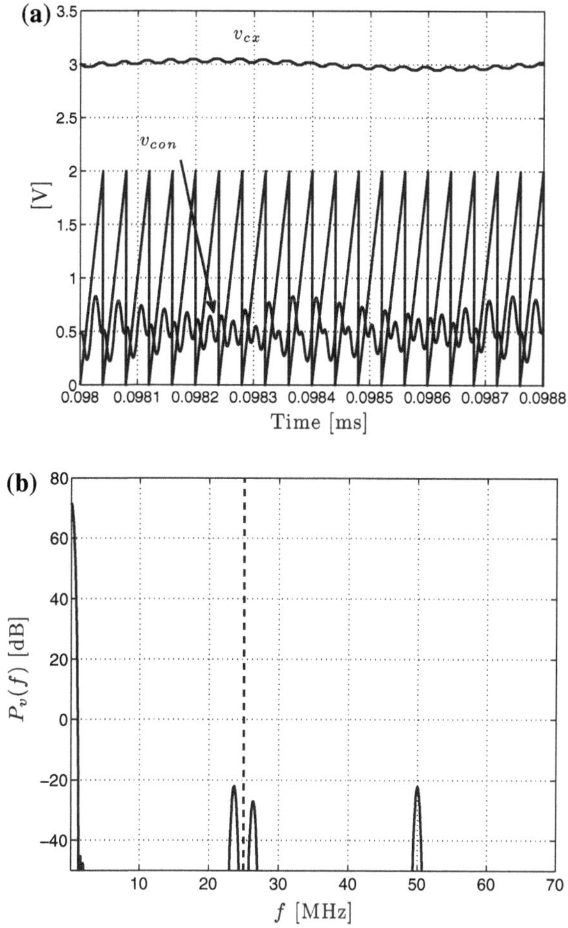

Fig. 5.3 Bifurcation diagram, sampled at the switching frequency f_s, by sweeping the proportional gain k_p for the multilevel converter with $C_x = 200$ nF **a** output voltage v_C **b** floating capacitor voltage v_{cx}

based characterization (and further analysis in the following sections) shows that decreasing the floating capacitance value C_x implies a reduction of the stability margin.

5.2 Experimental Results

This section explores FSI behavior and stability margins in an experimental prototype of a three-level buck converter. While in the previous section the component values were oriented to on-chip integration, in this section component values have been

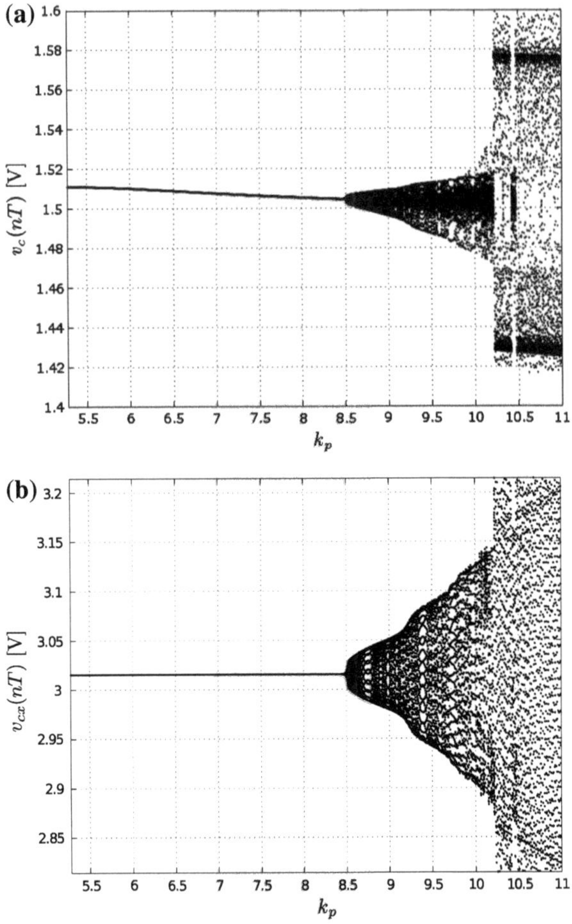

Fig. 5.4 **a** Time-domain waveform of VMC three-level control voltage v_{con}, floating capacitor voltage v_{cx} and ramp **b** Output waveform spectrum with $k_p = 9$ and $C_x = 200\,\text{nF}$

chosen to facilitate the implementation of the prototype, namely $V_g = 10\,\text{V}$, $V_{ref} = 2.5\,\text{V}$, $V_m = 0.935\,\text{V}$, $f_s = 50\,\text{kHz}$, $R = 2.5\,\Omega$, $k_p = 4$, $L = 33\,\mu\text{H}$, $C = 10\,\mu\text{F}$, $C_x = 10\,\mu\text{F}$ and $\omega_{z1} = 1$ krad/s.

Figure 5.6 shows the time and frequency domain experimental representation of the three-level converter before and after the instability is exhibited. The spectra show the existence of subharmonics at half of the switching frequency, even before instability is exhibited. This is not related to any instability but to the mismatching between both control phases in the multilevel converter, leading to slightly unbalanced floating capacitor voltage and then exhibition of small topology-inherent period-doubling. As it can be observed in the spectra domain, this undesired harmonic is negligible

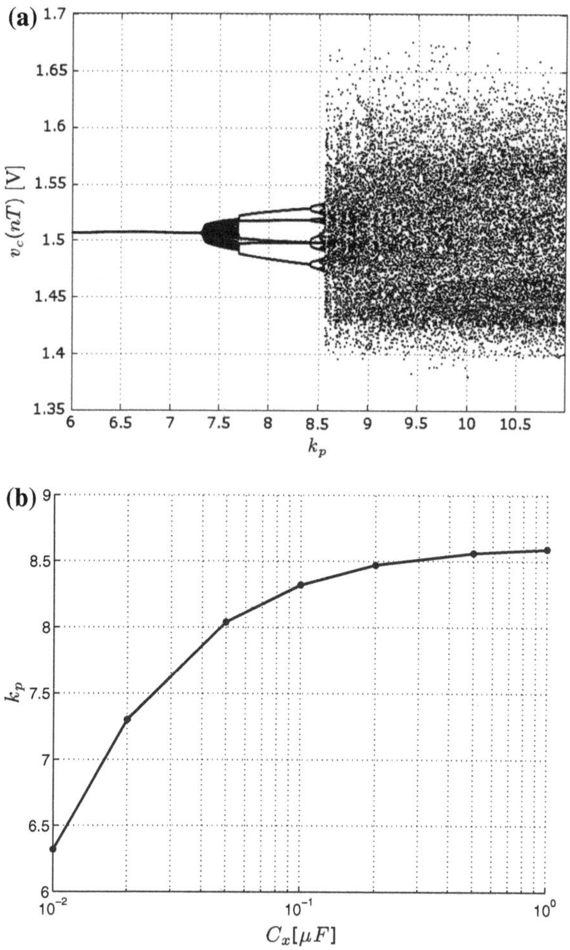

Fig. 5.5 **a** Bifurcation diagram by sweeping k_p and $C_x = 20\,\text{nF}$ **b** Stability boundary curve as a function of the proportional gain k_p and the floating capacitance C_x (semilog x-axis)

compared with the first harmonic at the switching frequency, whereas it is comparable when instability occurs.

Furthermore, this section experimentally characterizes the effect of the floating capacitance value C_x upon the stability boundary, as it is shown in Fig. 5.7. It is possible to observe that the lower C_x is, the smaller the critical proportional gain becomes, which agrees with the simulation results in Fig. 5.5.

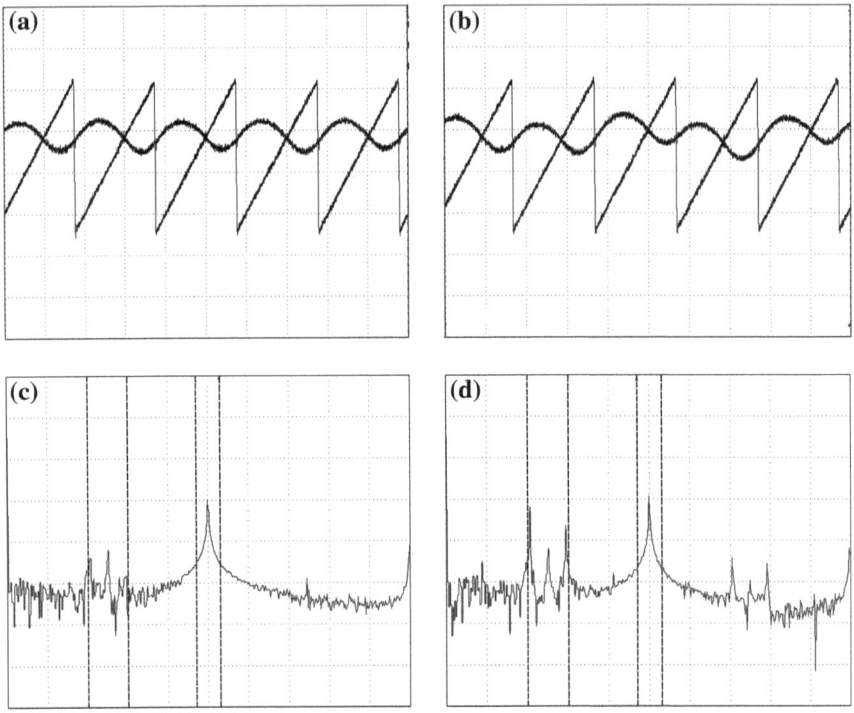

Fig. 5.6 Time-domain waveform and spectra representation from an experimental prototype before and after FSI takes place for $C_x = 10\,\mu\text{F}$

5.3 Three-Level Buck Converter Discrete-Time Model for FSI Prediction

In a similar procedure as it has been carried out in Appendix A for the conventional buck-based configuration cases, this section develops the discrete-time model for a three-level converter, shown in Fig. 5.8, which is composed of four possible topology configurations, and the whole switching cycle ($T_{cycle} = 2\,\text{T}$) is composed of two consecutive switching periods ($T = f_s^{-1}$), each of them composed by two configurations:

$$\dot{\mathbf{x}}_i = \mathbf{A}_{1,i}\mathbf{x}_i + \mathbf{B}_{1,i} \text{ for } t \in [iT, iT + d_i] \text{ for } i = 1, 2$$

where d_i is the fraction of time (duty cycle) during which the first configuration ($A_{1,i}$) is applied.

The three-level discrete-time model $\mathbf{P}$ that relates consecutive samples, namely the state variables at the beginning of the cycle $\mathbf{x}_n$ to those at the end of the same cycle $\mathbf{x}_{n+1}$, can be expressed as a composed function of two switching periods:

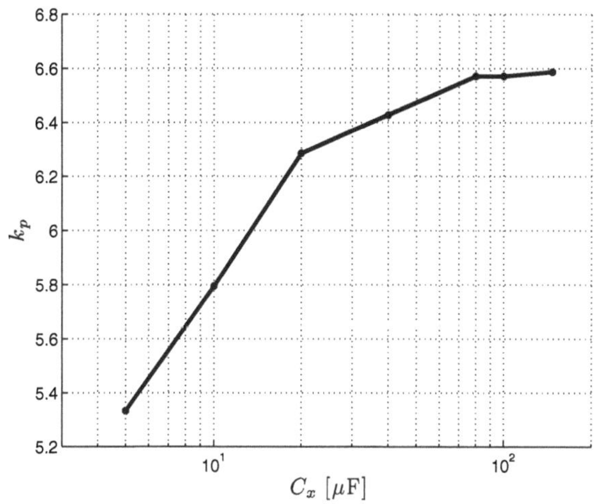

Fig. 5.7 Stability boundary curve as a function of the proportional gain k_p and the floating capacitance value C_x (semilog x-axis)

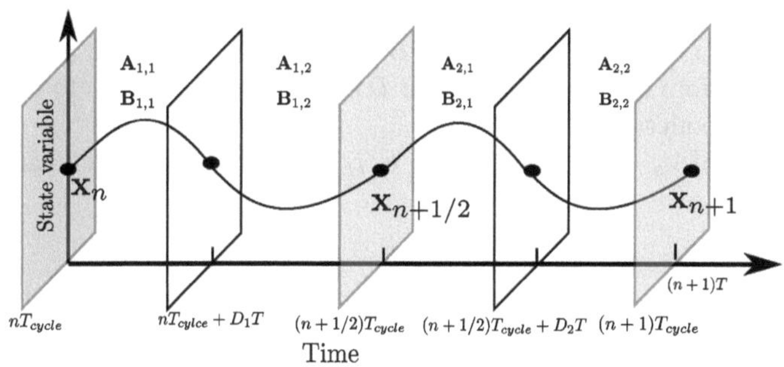

Fig. 5.8 Discrete-time model definition of a three-level buck converter

$$\mathbf{x}_{n+1} = \mathbf{P}_2(\mathbf{P}_1(\mathbf{x}_n)) \tag{5.1}$$

where,

$$\mathbf{P}_i = \Phi_i(d_i)\mathbf{x}_i + \Psi_i(d_i) \text{ for } i = 1, 2 \tag{5.2}$$

and $\Phi_i(d_i) = \Phi_{2,i}(\bar{d}_i T)\Phi_{1,i}(d_i T)$, being $\bar{d}_i = 1 - d_i$ and

$$\Phi_{1,i}(t_k) = e^{\mathbf{A}_{1,i}t_k},\ \Phi_{2,i}(t_k) = e^{\mathbf{A}_{2,i}t_k} \tag{5.3}$$

$$\Psi_i(d_i) = \Phi_{2,i}(\bar{d}_i T)\Psi_{1,i}(d_i T) + \Psi_{2,i}(\bar{d}_i T) \tag{5.4}$$

being $\Psi_{m,i}(t_k) = \int_0^{t_k} e^{\mathbf{A}_{m,i}\tau} d\tau \mathbf{B}_{m,i}$.

Additionally, there are two switching conditions σ_i, which depend upon the control law and the saw-tooth PWM modulator ramp signal $h(t)$:

$$\sigma_i(d_i T) = \mathbf{K}(\mathbf{X}_{ref} - \Phi_{1,i}(d_i T)\mathbf{x}_i) - h_i(d_i T) \text{ for } i = 1, 2 \tag{5.5}$$

being $\mathbf{K}$ the feedback gain vector corresponding to each state variable and having $h_1(t)$ and $h_2(t)$ a phase shift of 180° as it is shown in Fig. 5.1.

The stability analysis is carried out by studying the local behaviour of the model in the vicinity of steady-state $\mathbf{x}^*$, thereby extracting a Jacobian matrix $\mathbf{D\overline{P}}$ using the chain rule:

$$\mathbf{D\overline{P}} = \mathbf{D\overline{P}}_2 \mathbf{D\overline{P}}_1 \tag{5.6}$$

where,

$$\mathbf{D\overline{P}}_i = \mathbf{J}_{x,i} - \mathbf{J}_{d,i} J_{\sigma_{d,i}}^{-1} \mathbf{J}_{\sigma_{x,i}} \tag{5.7}$$

Each one of these terms can be expressed as:

$$\mathbf{J}_{x,i} = \Phi_{2,i}\Phi_{1,i}, \mathbf{J}_{d,i} = \Phi_{2,i}\Delta\dot{\mathbf{x}}_i \tag{5.8}$$

$$J_{\sigma_{d,i}}^{-1} = -\mathbf{K}\dot{\mathbf{x}}_i(DT^-) - m_c, \mathbf{J}_{\sigma_{x,i}} = -\mathbf{K}\Phi_{1,i}(DT) \tag{5.9}$$

In the previous equations, it is assumed that, in steady state, $d_n = D, \overline{D} = 1 - D$ and $\Delta\dot{\mathbf{x}}_i = \dot{\mathbf{x}}_i(DT^-) - \dot{\mathbf{x}}_i(DT^+)$, being $\dot{\mathbf{x}}_i(DT^-)$ the time derivative, in steady-state $\mathbf{x}_i = \mathbf{x}^*$, just before the switching instant DT, and m_c corresponds to the slope of the PWM saw-tooth ramp signal.

For the case of $V_o < V_g/2$ and $\mathbf{x}^T = \begin{bmatrix} v_c \; i_L \; v_{cx} \end{bmatrix}$, the state matrices are:

$$\mathbf{A}_{1,1} = \begin{pmatrix} \frac{-1}{RC} & \frac{1}{C} & 0 \\ \frac{-1}{L} & 0 & \frac{-1}{L} \\ 0 & \frac{1}{C_x} & 0 \end{pmatrix}, \quad \mathbf{A}_{1,2} = \begin{pmatrix} \frac{-1}{RC} & \frac{1}{C} & 0 \\ \frac{-1}{L} & 0 & \frac{1}{L} \\ 0 & \frac{-1}{C_x} & 0 \end{pmatrix} \tag{5.10}$$

$$\mathbf{A}_{2,1} = \mathbf{A}_{2,2} = \begin{pmatrix} \frac{-1}{RC} & \frac{1}{C} & 0 \\ \frac{-1}{L} & 0 & \frac{1}{L} \\ 0 & 0 & 0 \end{pmatrix} \tag{5.11}$$

$$\mathbf{B}_{1,1} = \begin{pmatrix} 0 \\ \frac{V_g}{L} \\ 0 \end{pmatrix}, \quad \mathbf{B}_{1,2} = \mathbf{B}_{2,1} = \mathbf{B}_{2,2} \begin{pmatrix} 0 \\ 0 \\ 0 \end{pmatrix} \tag{5.12}$$

The family of root loci obtained by sweeping the feedback gain k_p and C_x is shown in Fig. 5.9. The eigenvalues cross the unit circle having an imaginary part

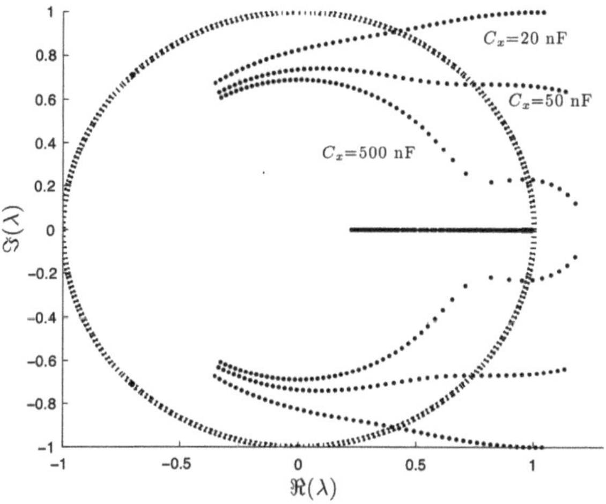

Fig. 5.9 Eigenvalues loci from the discrete-time model of VMC three-level buck converter by sweeping k_p and C_x

different from 0 due to the effect of the floating capacitance C_x, which agrees with the aforementioned exhibition of SSI. Note that even for very high values of C_x, the eigenvalues never cross the unit circle from 1.

Despite its powerful instability prediction capability, this approach requires complex mathematical tools which yields a map the parameters of which can be swept for stability characterization, but lacks a circuit-centric design-oriented perspective.

5.4 Ripple-Based Design-Oriented Index for Fast-Scale Instability Prediction

In this section, the ripple-based index condition presented in Chap. 3 is proposed for predicting the occurrence of the bifurcation boundary for a three-level buck converter, thereby demonstrating the general-purpose applicability of this approach. Note that, despite in the previous chapter the ripple-based index has been used to predict FSI boundary, both systems stability boundaries should be equivalent since in the case of a very large floating capacitance value, both the three-level buck converter (3L) and the elementary buck converter are equivalent save the voltage applied to the low-pass filter.

$$\rho_{3\text{L}} := \frac{k_p}{V_m}\hat{V}_{C,3\text{L}} < \rho_{crit,3\text{L}} \tag{5.13}$$

where the three-level buck converter ripple $\hat{V}_{C,3L}$, can be approximated by the same expression as in a conventional buck converter given in Eq. (3.11), but the battery voltage divided by 2:

$$\hat{V}_{C,3\mathrm{L}} := \frac{V_g D \bar{D}}{16 L C f_s^2} \tag{5.14}$$

Figure 5.10 shows the bifurcation diagram by sweeping various parameters. In all cases, there is an exhibition of SSI, which eventually leads to chaos. In this figure, it can also be observed that, by obtaining the bifurcation diagram as a function of their associated ripple-based index ρ, the bifurcation boundary remains almost constant, independently of the swept parameter.

Further characterization in Fig. 5.11 shows that ρ_{crit} varies with the voltage conversion ratio, in accordance with previous results for the conventional buck converter. The figure contrasts results for a three-level VCM buck converter with those for a conventional (two-level, 2L) VCM buck converter, demonstrating that due to its inherent lower ripple, and depending on the voltage conversion ratio, the three-level converter can have a wider stability range. It can be observed that, for $V_{ref}/V_g \simeq 0.25$, the three-level converter is further away from the stability boundary ($\rho_{3\mathrm{L}} < \rho_{crit,3\mathrm{L}}$), whereas close to $V_{ref}/V_g \simeq 0.5$, the conventional buck converter is further away from FSI boundary ($\rho_{2\mathrm{L}} < \rho_{crit,2\mathrm{L}}$).

More validation of the ripple-based approach has been carried out in Fig. 5.12 by contrasting the outcome of the discrete-time model, from Sect. 5.3, with results from the ripple condition (Eq. (5.13)). Regardless of the fact that the error increases for low values of the output resistor R and low values of the inductance L, which also occurs in the ripple-based index validation for the conventional buck converter in Chap. 3, the low approximation error in the rest of the design-space allows to conclude the appropriateness of this design-oriented approach for predicting FSI. Note that the explored portion of the design-space as a function of these parameters is limited by the DCM.

5.4.1 Effect of Low Floating Capacitor and Integration Limits

As discussed previously, it is of especial interest to explore the effect of low C_x upon the stability. The availability of the ripple-based approach, which joins the effect of different parameters into a single index, facilitates comparing within the complete design-space, the advantages and disadvantages in terms of stability for both topologies.

The effect of C_x upon ρ_{crit} is shown in Fig. 5.13, clearly reducing the stability boundary for low values of C_x.

This behavior can entail that the three-level converter can be more prone than the conventional buck converter to exhibit instability in the cases of very low values of C_x. This critical value of the floating capacitance C_x boundary has been explored in Fig. 5.14 under the best stability case condition, namely $V_{ref}/V_g \simeq 0.25$, showing

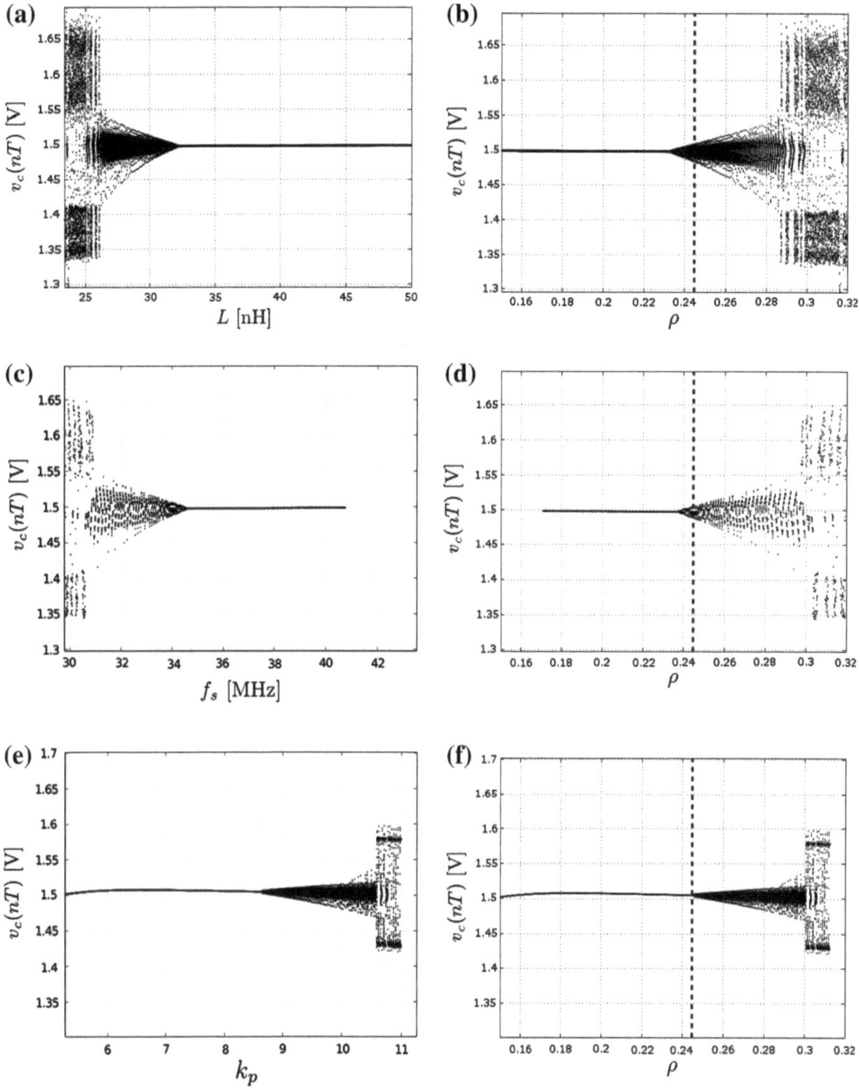

Fig. 5.10 Bifurcation diagram obtained by sweeping parameters **a** L, **c** f_s and **e** k_p and its equivalent representation as a function of the ripple-index ρ_{3L} **b**, **d** and **f**, respectively. $\rho_{crit,3L} = 0.245$ (*dashed-line*) with $C_x = 1\,\mu F$

that for low values of the floating capacitance value C_x, the three-level has worse stability boundary than a conventional buck converter.

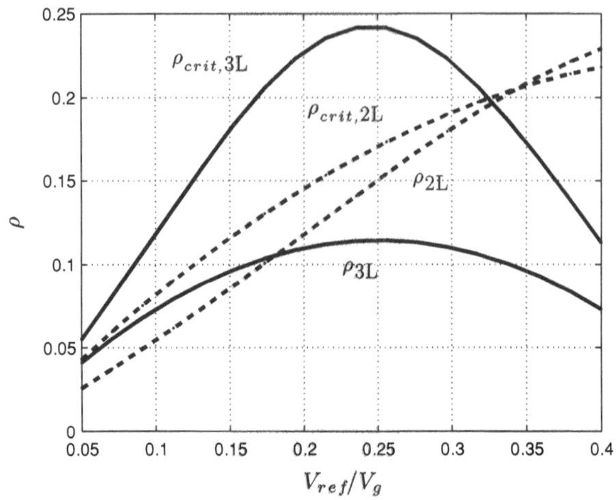

Fig. 5.11 Stability boundary ρ_{crit} and ripple-based index ρ for a three-level VMC buck converter (3L, *solid*) and for an conventional buck converter (2L, *dashed line*) as a function of voltage conversion ratio $V_{ref}\,/\,V_g$

5.5 Buck-Based Switching Power Amplifier: Modeling and Characterization

The work in the field of FSI analysis in switching power converters has hitherto been mainly focused on regulation applications. The work in this section focuses instead on the effect of the reference dynamics (this is, a time-varying reference tracking application), upon the FSI boundary (Fig. 5.15).

A classification of representative cases of tracking applications could include four categories, namely (a) Quasistatic variation of the reference voltage (b) Low-frequency sinusoidal reference variation, (c) High-frequency sinusoidal reference, close to filter and switching frequency and (d) Noise-like wideband signal. Although the latter corresponds to actual applications, this work focuses on the three first simple but representative cases. The parameter values considered in this section are: $V_g = 6\,\text{V}$, $V_{ref} = 3\,\text{V}$, $V_m = 1\,\text{V}$, $f_s = 50\,\text{MHz}$, $R = 2.5\,\Omega$, $k_p = 3$, $L = 66\,\text{nH}$, $C = 20\,\text{nF}$, $\omega_{z1} = 0.1\,\text{Mrad/s}$.

The simplest case consists of a quasi-static variation of the output voltage level of the converter depending on system-level power requirements. Representative application-level examples include a slow case of dynamic voltage scaling DVS for supplying microprocessors depending on computing workload, and quasistatic system-level control of supply voltage in RFPAs. Therefore its stability will depend on the output level required in each instant. Figure 5.16 shows the output voltage for different reference voltage levels by maintaining converter and feedback parameters. Note that in case of high or low voltage levels, the converter is more prone to exhibit instability.

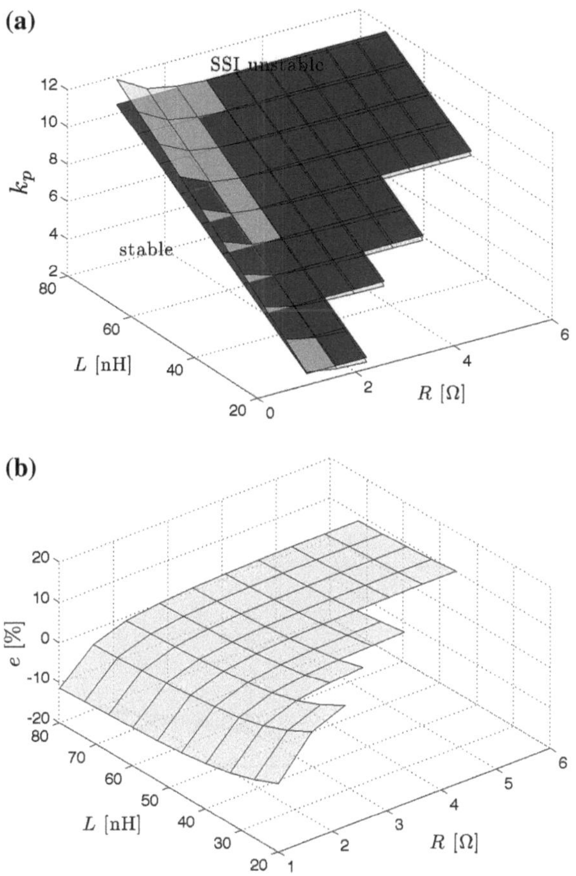

Fig. 5.12 Stability boundary surface obtained from the discrete-time model (*black*) and ripple-based index condition (*white*), given in Eq. (5.13) with ρ_{crit}=0.245 as a function of **a** inductance L, output resistance R and the proportional gain k_p

An example that is conceptually similar to case (b) of a quasistatic dynamic reference in power switching power converter is presented in (Tse et al. 2003a), (Wu et al. 2005), (Li et al. 2008), where the stability of power factor correction (PFC) is evaluated by a reference sinusoidal supply voltage, the frequency of which is very low (~50 Hz).

This work focuses on the analysis of the stability boundary for the nonlinear switched system in the context of high frequency tracking applications, representative of audio amplifiers, envelope trackers in polar RF transmitters and line drivers for Power Line Communications (PLC). In these cases, the reference has a wide-band characteristic (the ratio of the switching frequency f_s to the modulation frequency f_m is relatively high) and its amplitude covers the converter output voltage range.

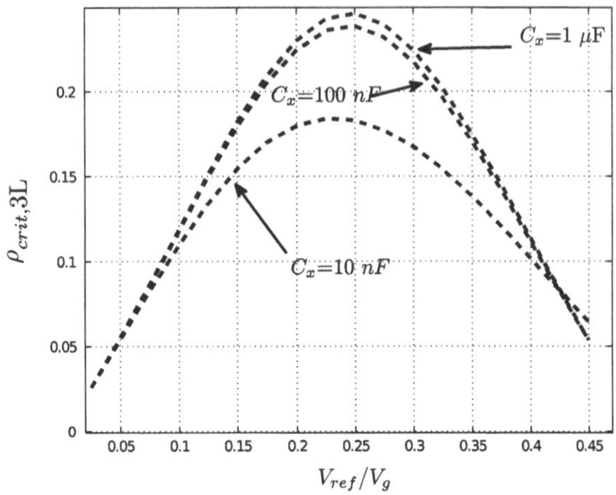

Fig. 5.13 Stability boundary $\rho_{crit,3L}$ as a function of voltage conversion ratio V_{ref}/V_g and the floating capacitance value C_x

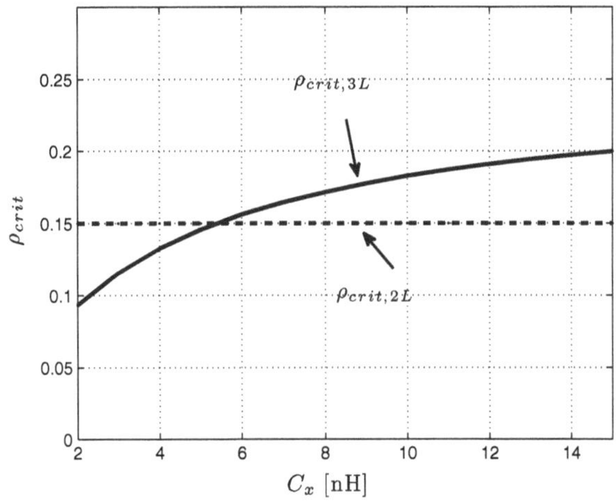

Fig. 5.14 Stability boundary ρ_{crit} as a function of floating capacitance C_x and $V_{ref} = 1.5\,\text{V}$

5.5.1 Buck-Based Switching Power Amplifier Instabilities: Design-Space Exploration

Switching power amplifiers demand a voltage time-varying reference, which should be tracked by the switched system, hence sophisticating the system with more complex dynamics. As a representative case of such wide band signal, a single tone is initially considered in this section:

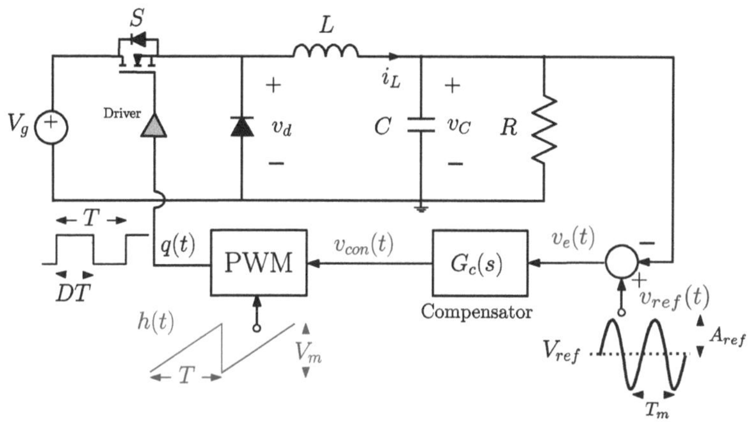

Fig. 5.15 VMC Buck converter with time-varying sinusoidal $V_{ref}(t)$ reference

$$v_{ref}(t) = V_{ref} + A_{ref}\sin(2\pi f_m t) \tag{5.15}$$

Adding an external reference extends the number of parameters in the design-space, as observed in Fig. 5.17, in which it is shown the frequency domain representation of the output voltage along with the frequency response of the LC filter of the buck converter. From such representation, the new design parameter space can be described with a compressed set of parameters, namely the ratio of the three frequencies in the system:

$$\Gamma_s = \frac{f_s}{f_0} \quad \Gamma_m = \frac{f_0}{f_m} \tag{5.16}$$

where $f_0 = 1/(2\pi\sqrt{LC})$. In addition, the reference amplitude A_{ref} should be considered also in the design parameter space, while V_{ref} (offset of the reference signal) will be considered constant in the next characterization and centered at half of the voltage source V_g to maximize the dynamic range. The nominal values in this section are: $A_{ref} = 0.5$ V and $f_m = 1$ MHz.

To initially illustrate the nonlinear dynamic behavior, Fig. 5.18 shows the output voltage waveform both when the switching amplifier is stable and when FSI is exhibited, this is, the feedback loop forces the output to track the sinusoidal reference on average, but period-one constant switching frequency operating is lost. It is complemented with a sampled map, obtained by sampling the output voltage synchronously to the PWM switching frequency f_s. The output voltage manifests FSI during some time interval of the waveform.

To comprehensively characterize the design-space, the first studied parameters are the proportional gain k_p, the inductance L, and the switching frequency f_s. Figure 5.19 shows a sampled waveform obtained by sampling the control waveform v_{con}. By reducing the inductance value or switching frequency, or increasing the

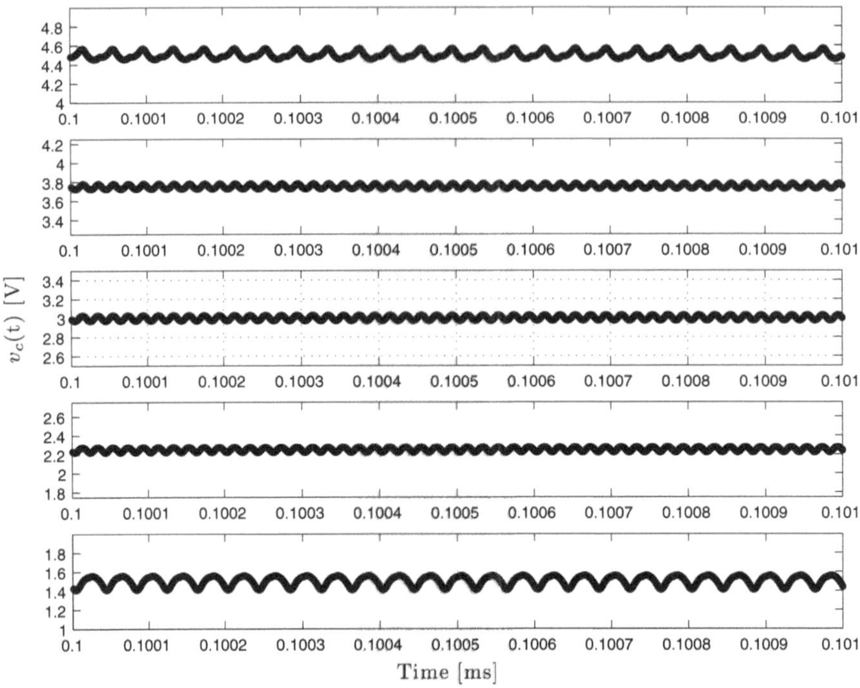

Fig. 5.16 Steady-state output voltage waveform for different voltage references in a buck converter. $V_{ref} = 1.5, 2.25, 3, 3.75, 4.5\,\mathrm{V}$

proportional gain, FSI is exhibited. Note that these trends are qualitatively the same as it has been observed in regulation application.

In Fig. 5.20, the load resistance has been considered, showing a slight impact upon FSI boundary (note that simulations are carried out very close to the stability boundary in order to be able to observe the effect of such resistance) in a similar way as in a static reference regulator application, in which the resistance has negligible effect upon FSI (except for low values).

To conclude this section, it is considered the effect of the modulation frequency f_m and modulation amplitude A_{ref} in Fig. 5.21. Note the singularity of both parameters since they are particular to signal tracking applications.

From this characterization, it is possible to observe that the higher the amplitude A_{ref} is, the more prone to exhibit FSI the system becomes, which is in agreement with what was observed in regulation applications, since this parameter directly implies increasing the duty cycle ranges. Conversely, the effect of the modulator frequency f_m is very small, even improving the system stability in terms of FSI.

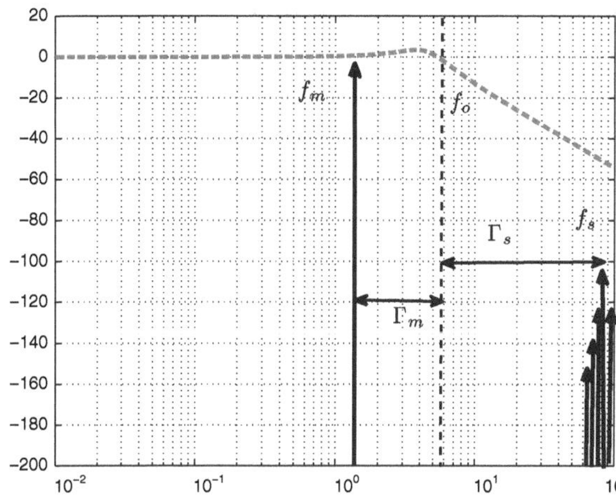

Fig. 5.17 Frequency representation of buck LC filter (*dashed*) and harmonic composition of PWM waveform (centered at f_s) modulated by a sinusoidal reference (at f_m)

5.5.2 *Qualitative Characterization of Instabilities in a Buck-Based Power Switching Amplifier with Time-Varying Sinusoidal Reference*

While in regulation applications the FSI produce an undesirable increase in the time-domain output voltage ripple, in power amplification applications the focus is on characterizing the effect of FSI upon the spectrum, since applications such as adaptive RF power amplifier supply are subject to strict spectral masks. Furthermore, the section also studies the effect of FSI upon the effective switching frequency of the converter. This parameter is of especial interest in wide-band power supplies since, due to the high bandwidth signal, the need of using higher switching frequency has an important detrimental impact on efficiency through switching losses. As it has been shown in Sect. 2.2 in regulation, chaotic behavior can decrease the average switching frequency hence improving the efficiency.

The evolution of the time and frequency domain behavior under different representative conditions is shown in Fig. 5.22 by sweeping the switching frequency f_s. The main interest of sweeping this parameter is to know the effect of reducing the ratio between the modulator and switching frequency.

The switching amplifier exhibits period-doubling behavior that ends up in a chaotic regime. The exhibition of such chaotic behavior produces low-frequency noise-like harmonics with a similar level as the switching frequency, limiting the potentials benefits of the spectral distribution carried out by chaotic behavior. Furthermore, as it can be observed in Fig. 5.23, this chaotic behavior has associated a time-domain multiple switching phenomenon hence increasing the average switching frequency. Note that after the chaotic behavior occurs, by continuously decreasing

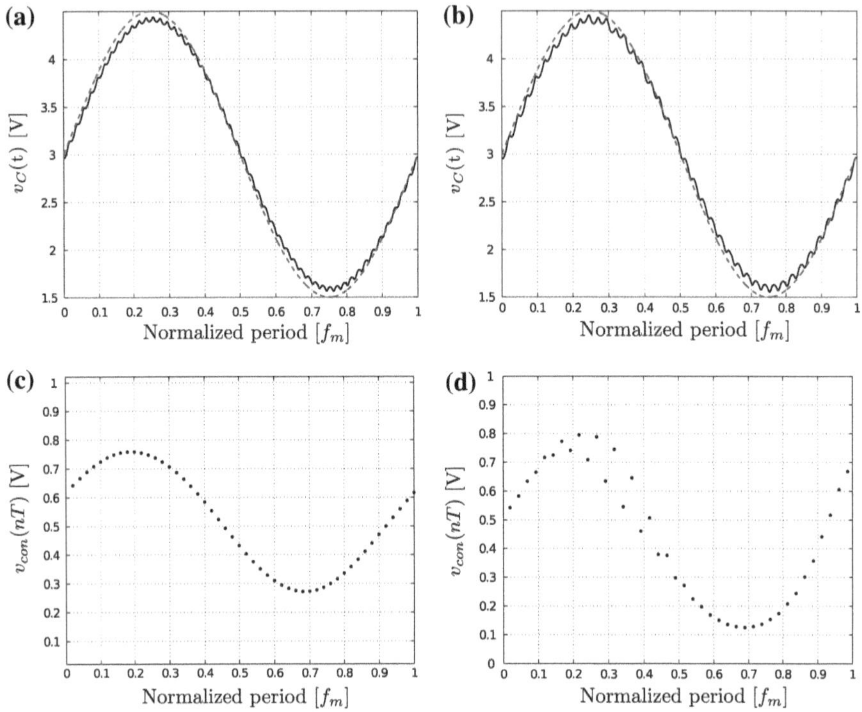

Fig. 5.18 Time-domain output voltage and reference (*dash*) for **a** $f_s = 50\,\text{MHz}$ **b** $f_s = 40\,\text{MHz}$. Associated sampled map at the switching frequency f_s in**c** and **d**, respectively

the switching frequency, a new period-one behavior window takes place, as it is shown in Fig. 5.24.

5.5.3 Description of the Switching Amplifier Dynamics by a Discrete-Time Model

Being characterized by two different forcing periods, two kinds of discrete-time modeling for the system can be obtained. If concerned with the dynamics of the system within the switching cycle, the first order discrete-time model can be defined from the current switching cycle to the next one. This is the mapping that relates the state variables from the beginning to the end of a switching cycle. If looking for the dynamics of the system during a reference cycle f_m then the second order discrete-time model is considered. Next, we will obtain the first order discrete-time model as an extension of what has been done in the Appendix A for regulation cases.

The system configuration during each switching sub-interval is linear and time-varying as it is mentioned earlier. During the ON phase the trajectory of the system,

Fig. 5.19 Sampled waveforms at the switching frequency for different values of **a** the proportional gain k_p with $k_p = 3$ (*dash-dot*) and $k_p = 6$ (*dots*), **b** the inductance value L with $L = 66\,\text{nH}$ (*dash-dot*) and $L = 44\,\text{nH}$ (*dots*) **c** and the switching frequency f_s with $f_s = 50$ and $40\,\text{MHz}$

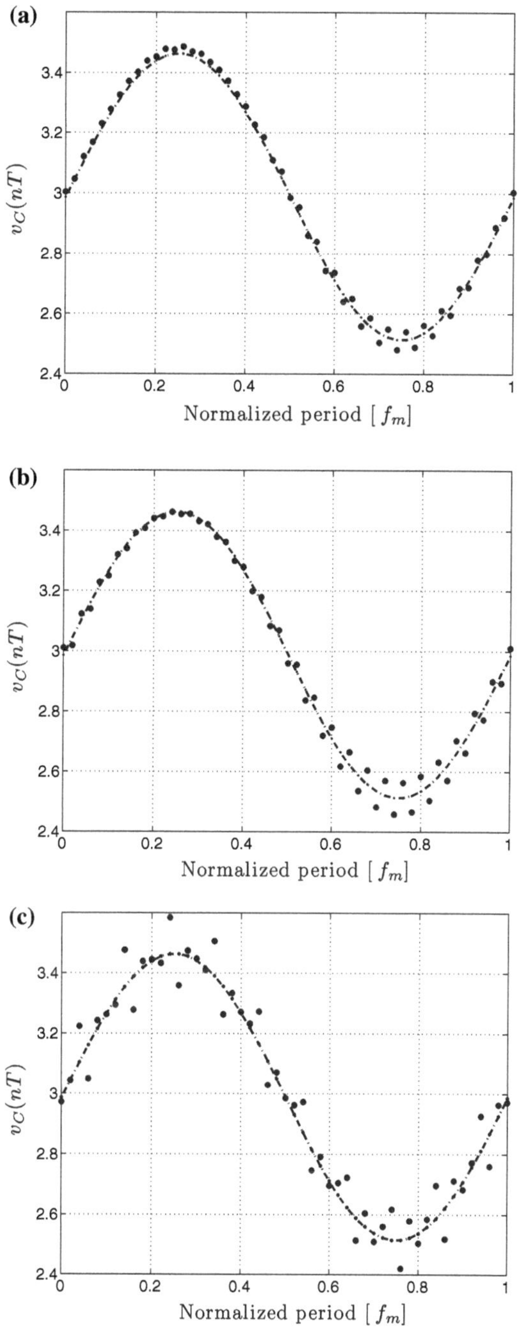

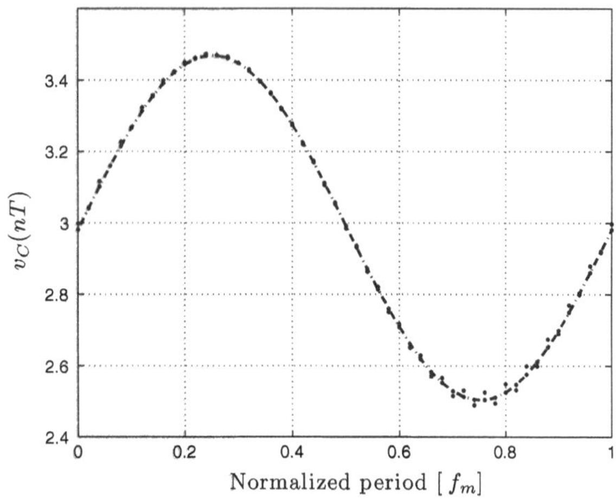

Fig. 5.20 Sampled waveforms at the switching frequency for different values of output resistor R with $R = 1\,\Omega$ (*dash-dot*) and $R = 2.5\,\Omega$ (*dots*) with $k_p = 4.5$

starting from the initial condition $\mathbf{x}_n$, is expressed by:

$$\mathbf{x}(t) = e^{\mathbf{A}(t-nT_s)}\mathbf{x}_n + \int_{nT_s}^{nT_s+t} e^{\mathbf{A}\tau}\mathbf{B}_1(\tau)d\tau \tag{5.17}$$

At the time instant d_nT, the system switches from ON phase to OFF phase. This instant can be determined from the equation resulting from the crossing of the ramp signal $v_{\text{ramp}}(t)$ with the control voltage $v_{con}(t)$. This equation can be written as follows:

$$\sigma(d_nT) = \mathbf{K}(\mathbf{X}_{ref}(d_nT) - \Phi_1(d_nT)\mathbf{x}_n) - h(d_nT) \tag{5.18}$$

where $\mathbf{X}_{ref}(t) = \left(V_{ref} + A_{ref}\sin(\omega_m t)\ 0\ 0\right)^T$ is a vector that depends upon time. At the instant $(n+1)T$ the system switches to ON phase again. The state of the system at time instant $(n+1)T_s$ is given by:

$$\mathbf{x}_{n+1} = e^{\mathbf{A}T_s}\mathbf{x}_n + e^{\mathbf{A}(1-d_n)T_s} \int_{nT_s}^{(n+d_n)T_s} e^{\mathbf{A}\tau}\mathbf{B}_1(\tau)d\tau + \int_{(n+d_n)T_s}^{(n+1)T_s} e^{\mathbf{A}\tau}\mathbf{B}_2(\tau)d\tau \tag{5.19}$$

It is worth noting here that (5.18) is a transcendental equation and that a root-finding algorithm must be applied in order to obtain the duty cycle for each switching period. In order to simplify the writing of the first order map, let us write it in the following form:

$$\mathbf{x}_{n+1} := \mathbf{P}(\mathbf{x}_n, n) = \Phi\mathbf{x}_n + \Phi_2\Psi_1(n) + \Psi_2(n) \tag{5.20}$$

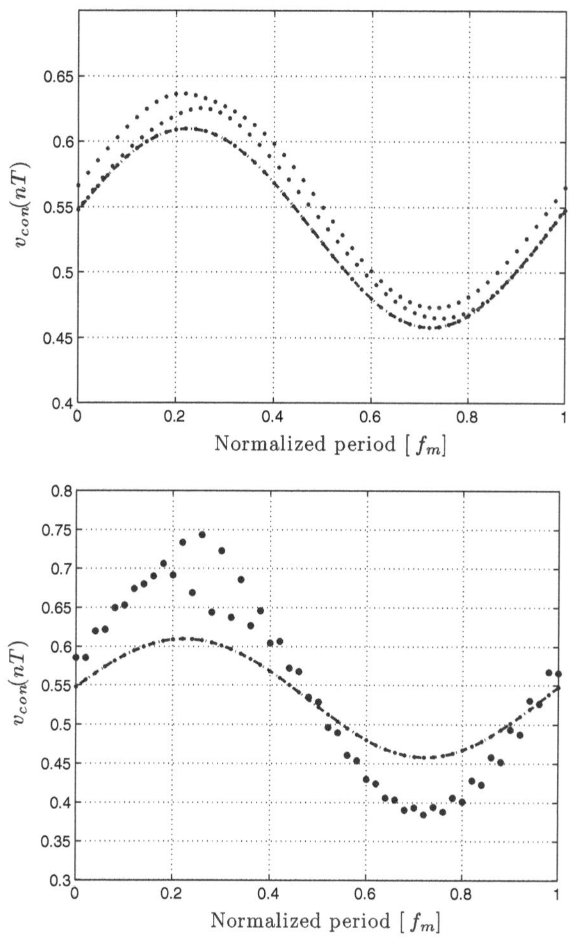

Fig. 5.21 Sampled waveforms at the switching frequency for different values of the modulator frequency f_m, with $f_m = 1$ MHz (*dash-dot*) and $f_m = 0.5$ MHz (*dots*) and the modulator amplitude A_{ref} with $A_{ref} = 0.5$ and 1 V. With $k_p = 4.3$

where

$$\Phi = e^{\mathbf{A}T}, \quad \Phi_1 = e^{\mathbf{A}d_nT}, \quad \Phi_2 = e^{\mathbf{A}((1-d_n)T)}$$
$$\Psi_1(n) = \int_{nT}^{(n+d_n)T} \Phi_1 \mathbf{B}_1(\tau)d\tau, \quad \Psi_2(n) = \int_{(n+d_n)T}^{(n+1)T} \Phi_2 \mathbf{B}_2(\tau)d\tau \tag{5.21}$$

Note that the vectors $\Psi_k(n)$ are time-dependent, making the first order Poincaré map time-varying. Once the matrices Φ_k are obtained, an expression for matrices $\Psi_k(n)$ can be found from Eq. (5.21). For simplicity of integration of the system equations during each phase, let us write the vectors $\mathbf{B}_1(t)$ and $\mathbf{B}_2(t)$ as the sum of a constant term $\mathbf{B}_a$ and a T_m−periodic time-varying term $\mathbf{B}_b \sin(\omega_m t)$:

$$\mathbf{B}_1(t) = \mathbf{B}_{a,1} + \mathbf{B}_b \sin(\omega_m t), \mathbf{B}_2(t) = \mathbf{B}_{a,2} + \mathbf{B}_b \sin(\omega_m t) \tag{5.22}$$

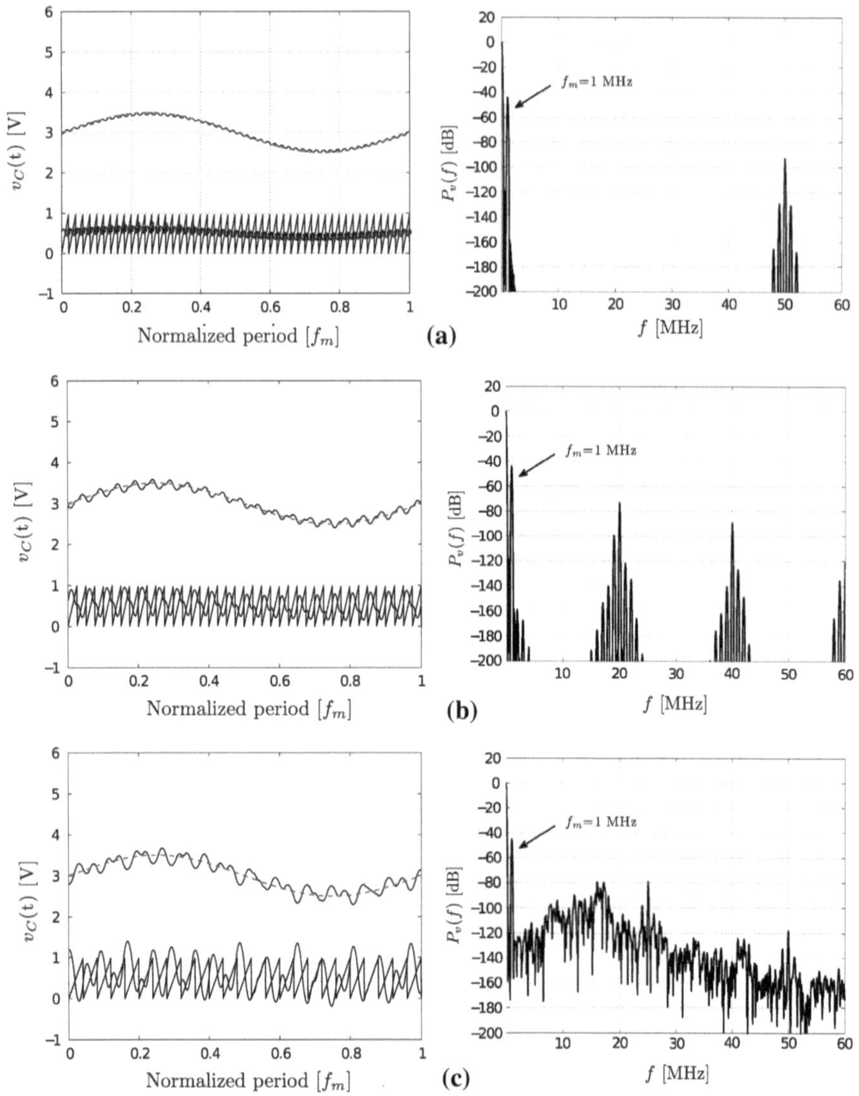

Fig. 5.22 Time domain (*left*) and spectra (*right*), being $P = 20\log(|V_c(f)|)$, with **a** $f_s = 50\,\text{MHz}$ **b** $f_s = 40\,\text{MHz}$ and **c** $f_s = 25\,\text{MHz}$

where

$$\mathbf{B}_{a,1} = \begin{pmatrix} 0 \\ \frac{V_g}{L} \\ V_{ref} \end{pmatrix},\ \mathbf{B}_{a,2} = \begin{pmatrix} 0 \\ 0 \\ V_{ref} \end{pmatrix} \text{ and } \mathbf{B}_b = \begin{pmatrix} 0 \\ 0 \\ V_{ref} \end{pmatrix} \tag{5.23}$$

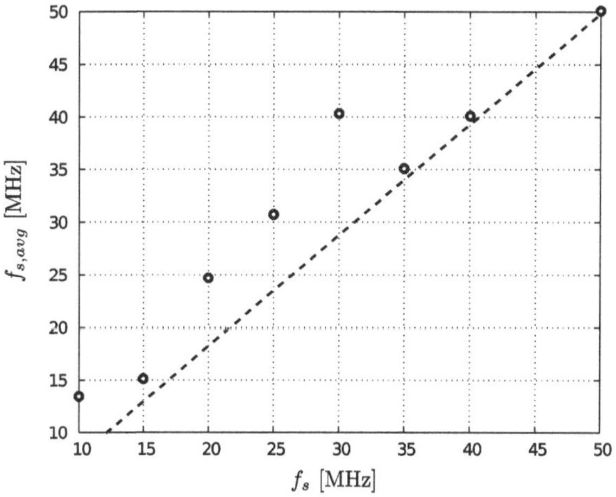

Fig. 5.23 Average switching frequency $f_{s,avg}$ versus the switching frequency f_s

By computing the integral term in (5.17), $\Psi_1(n)$ can be expressed as:

$$\begin{aligned}\Psi_1(n) = &(\mathbf{A}^2 + \omega_m^2\mathbf{I})^{-1}[\omega_m\Phi_1\mathbf{B}_b\cos(n\omega_m T_s) \\ &+ \Phi_1\mathbf{A}\mathbf{B}_b\sin(n\omega_m T_s) - \omega_m\mathbf{B}_b\cos((n+d_n)\omega_m Ts) \\ &+ \mathbf{A}\mathbf{B}_b\sin((n+d_n)\omega_m T_s)] + \int_{nT}^{(n+d_n)T} e^{\mathbf{A}\tau}d\tau\mathbf{B}_{a,1}\end{aligned} \tag{5.24}$$

The vector $\Psi_2(n)$ can be obtained in the same way. Note that different expressions for Ψ_k $(k = 1, 2)$ would be obtained if the reference voltage would be considered constant (quasi-static approximation). Note also that the function Ψ_k corresponding to a constant reference voltage can be obtained by forcing $f_m = 0$ in Eq. (5.24). Once we get both of Φ_k and $\Psi_k(n)$ and combining with Eq. (5.18), the expression of **P** is obtained from Eq. (5.20). **P** is the mapping that relates the vector of the state variables $\mathbf{x}_n$ at the beginning of the switching cycle to $\mathbf{x}_{n+1}$, those at the end of the same cycle. Note that it is a nonlinear map in the state variables and periodically time-varying in the discrete-time domain. This nonlinear time-varying map has a periodic orbit (not a fixed point) as a nominal operating regime. The period N of this discrete-time periodic orbit is Γ_m/Γ_s if this ratio is an integer number. If Γ_m/Γ_s is a rational number, the period will be the denominator of this number. If the periodic orbit is stable, and starting from a point $\mathbf{x}_0$ belonging to this periodic orbit, the map will take entirely N periods to return to the same point.

Numerical simulations are carried out using this map in a MATLAB code and an iterating procedure to generate the discrete-time waveforms. The results are obtained by using the same parameter values for their corresponding time-domain simulations from the switched model. Simulations corresponding to these cases are shown in

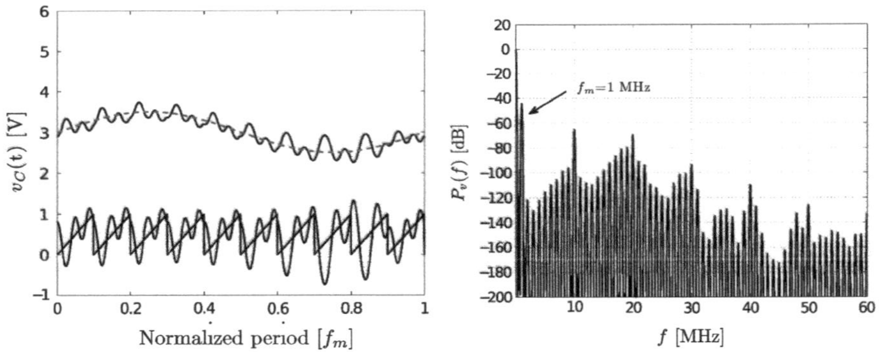

Fig. 5.24 Time domain (*left*) and spectra (*right*), being $P_v(f) = 20\log(|V_c(f)|)$, with **a** $f_s = 10\,\text{MHz}$

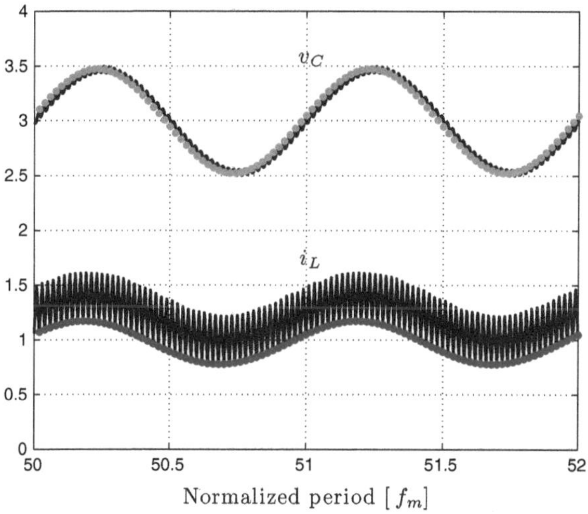

Fig. 5.25 Voltage and current waveforms of a buck-based switching amplifier from the discrete-time model (*dots*) and state equations numerical simulations showing normal periodic stable behavior with $f_s = 50\,\text{MHz}$

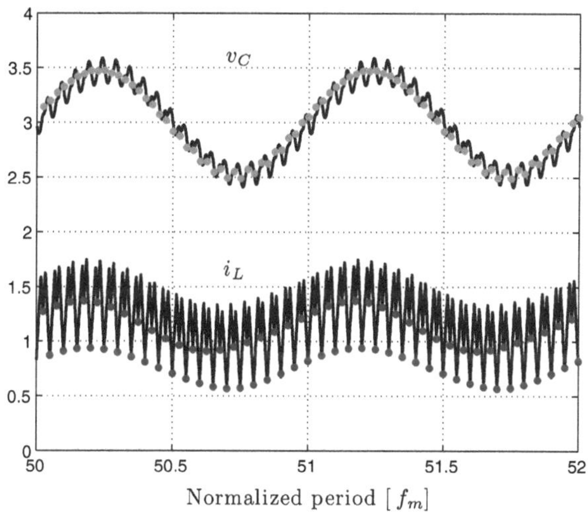

Fig. 5.26 Voltage and current waveforms of a buck-based switching amplifier from the discrete-time model (*dots*) and state equations numerical simulation showing period doubling bifurcation with $f_s = 40\,\text{MHz}$

Fig. 5.25 and 5.26. Figure 5.25 corresponds to the stable operation of the system while Fig. 5.26 corresponds to the s subharmonic oscillation at the switching period resulting from a period-doubling bifurcation. As it can be observed, there is a perfect agreement between the results obtained from the numerically integrated switched model and the derived analytical discrete-time model. However, the results obtained from the latter model are obtained much faster than those obtained by using the switched model as it is always more easy to iterate a recurrence equation than to numerically integrate a differential equation.

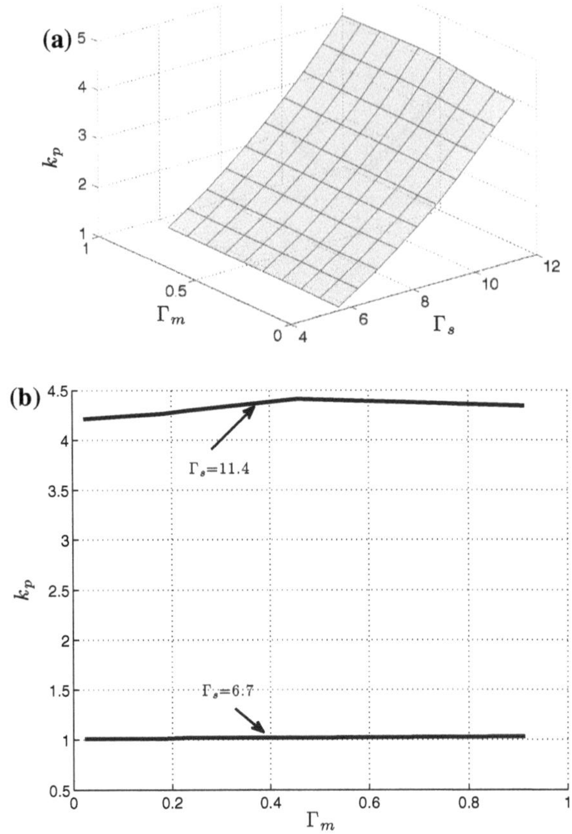

Fig. 5.27 **a** Stability boundary as a function of parameters Γ_m, Γ_s, and the proportional gain k_p. **b** 1-D cut as a function of Γ_m and k_p

5.5.4 Characterizing the Stability Boundary from the Discrete Time Model

In this section we will obtain the useful region (region of stable periodic behavior) of the buck-based switching amplifier in the parameter space.

As design parameters we will now choose Γ_m, Γ_s in addition to the gain of the PI controller k_p and the reference amplitude A_{ref}.

We vary Γ_m and Γ_s in a suitable grid and for each point in this grid we increase the value of k_p till obtaining a bifurcation. The results are shown in Fig. 5.27, obtained from the discrete-time model due to its faster processing time. Note that the reference frequency f_m is swept from low frequencies to the resonance frequency f_o since higher frequencies will result in inadequate tracking on average. Regarding the stability boundary variation with Γ_m, it is worth noting the slight stability improvement by increasing this parameter. Therefore, low frequencies, in which the design-oriented

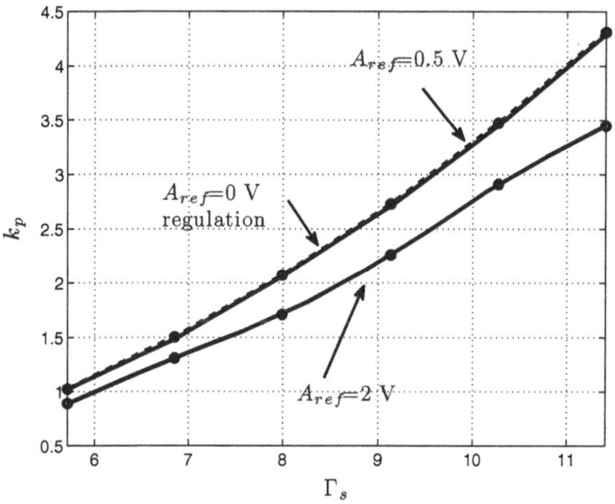

Fig. 5.28 Stability boundary as a function of Γ_s and k_p by using different reference amplitude, A_{ref} (V_{ref} = 3 V)

Fig. 5.29 Stability boundary as a function of A_{ref} and k_p by using modulation frequency f_m = 1 MHz and f_m = 0 (regulation), for a constant voltage reference V_{ref} = 3 − A_{ref} V

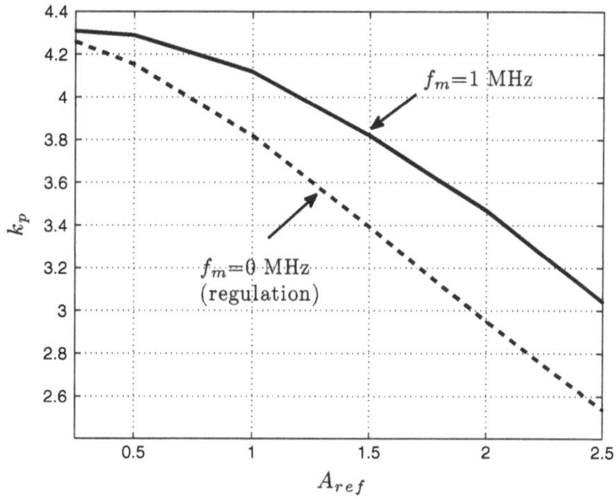

index ripple approach without consideting the Γ_m is valid, could be considered a conservative limit that guarantees that at higher reference frequencies, the output is stable even near the resonance frequency.

The simulations also show a higher influence of the parameter Γ_s compared to that of the modulating frequency. Figure 5.28 shows the variation of the maxium k_p as a function of Γ_s and compares it with the dependency in regulation application, almost showing similar trends. However, the stability curve changes when the reference of the amplitude is increased.

A complete study of the impact of the reference amplitude has been carried out in Fig. 5.29. Note that its influence upon stability agrees with what has been observed in the regulation case, for which stability boundary has a strong dependency upon duty cycle. Therefore, it is also compared with the stability limit obtained by sweeping such reference amplitude and its equivalent stability boundary in regulation. From such comparison, it is possible to observe that the regulation stability boundary, indeed, is a conservative limit.

Finally, and after having explored the design-space extended for the switching amplifier by considering the reference amplitude A_{ref} and frequency f_m, it is possible to conclude that under a sinusoidal external reference, by guaranteeing the FSI in regulation, it is also possible to ensure it in sinusoidal tracking.

Appendix A
Definition of the Discrete-Time Model

In Chap. 3 (Sect. 3.3), the discrete-time models have been developed for a general case, and afterwards particularized for a VMC buck converter under a proportional control. Indeed, particularizing for other kinds of control requires to change:

- the number of state-variables
- the state matrices
- the feedback gain vector, depending of which state-variables are added to the feedback loop

In the following subsections the discrete-time models for the systems considered in this book are derived.

A.1 PI Controller in a VMC or CMC Buck Converter

The s-domain error-to-control (e-to-v_{con}) PI controller transfer function is:

$$G_{c,PI}(s) := \frac{v_{con}(s)}{v_e(s)} = k_p \frac{s + \omega_{z1}}{s} \tag{A.1}$$

where $v_e(s)$ and $v_{con}(s)$ are the Laplace transform of the error signal $v_e = V_{ref} - v_C$ and the control signal v_{con} respectively. Equivalently, the control signal is given by the following time-domain expression

$$v_{con}(t) = k_p v_e(t) + k_I \int_{-\infty}^{t} v_e(\theta) d\theta \tag{A.2}$$

Therefore, the state-variable vector is $\mathbf{x} = (v_C i_L v_I)^T$, where $v_I = \int_{-\infty}^{t} (V_{ref} - v_c(\theta)) d\theta$ and the system state matrices are:

E. Rodríguez Vilamitjana et al., *Chaos in Switching Converters for Power Management*,
DOI: 10.1007/978-1-4614-2128-3,

$$\mathbf{A}_1 = \mathbf{A}_2 = \begin{pmatrix} \frac{-1}{RC} & \frac{1}{C} & 0 \\ \frac{-1}{L} & 0 & 0 \\ 1 & 0 & 0 \end{pmatrix}, \quad \mathbf{B}_1 = \begin{pmatrix} 0 \\ \frac{V_g}{L} \\ V_{ref} \end{pmatrix}, \quad \mathbf{B}_2 = \begin{pmatrix} 0 \\ 0 \\ V_{ref} \end{pmatrix} \tag{A.3}$$

The associated vector gain is:

$$\mathbf{K} = \begin{bmatrix} k_v & k_i & k_I \end{bmatrix} \tag{A.4}$$

where $k_I = k_p\omega_{z1}$, k_i is the current gain for the CMC case and $k_v = k_p$.

Note that by adjusting the voltage or current gains by modyifing k_v and k_i respectively, the converter can be a CMC or VMC. Equivalently to a CMC PI controller, the PID compensator case can be also written as a function of voltage and current gains as it has been carried out in Sect. 3.7.1.

A.2 Realistic PID Controller in a VMC Buck Converter

The transfer function of a realistic PID is:

$$G_{c,PID}(s) = G\frac{(s+\omega_{z1})(s+\omega_{z2})}{s(s+\omega_{p2})} \tag{A.5}$$

The system state matrices, including an additional order due to the new pole ω_{p2}, are:

$$\mathbf{A}_1 = \mathbf{A}_2 = \begin{pmatrix} \frac{-1}{RC} & \frac{1}{C} & 0 & 0 \\ \frac{-1}{L} & 0 & 0 & 0 \\ -k_I/\omega_{z2} & 0 & 0 & 0 \\ -k_v & -k_i & \omega_{p2} & \omega_{p2} \end{pmatrix}, \tag{A.6}$$

$$\mathbf{B}_1 = \begin{pmatrix} 0 \\ \frac{V_g}{L} \\ \frac{k_I}{\omega_{p2}} V_{ref} \\ k_p V_{ref} \end{pmatrix}, \quad \mathbf{B}_2 = \begin{pmatrix} 0 \\ 0 \\ \frac{k_I}{\omega_{p2}} V_{ref} \\ k_p V_{ref} \end{pmatrix} \tag{A.7}$$

where, k_v, k_i and k_I are the same as in the PID. The feedback gains vector can be expressed as:

$$\mathbf{K} = \begin{bmatrix} 0 & 0 & 0 & 1 \end{bmatrix} \tag{A.8}$$

A.3 DCM in a VMC Buck Converter Under PI Controller

The DCM implies an additional configuration during certain periods of time in which no current flows through the inductor. The system state matrices are given by:

$$\dot{\mathbf{x}} = \mathbf{A}_1\mathbf{x} + \mathbf{B}_1 \text{ for } t \in [nT, nT + d_nT]$$
$$\dot{\mathbf{x}} = \mathbf{A}_2\mathbf{x} + \mathbf{B}_2 \text{ for } t \in [nT + d_n, nT + d_{n,2}T]$$
$$\dot{\mathbf{x}} = \mathbf{A}_3\mathbf{x} + \mathbf{B}_3 \text{ for } t \in [nT + d_{n,2}T, nT + T]$$

where

$$\mathbf{A}_1 = \mathbf{A}_2 = \begin{pmatrix} \frac{-1}{RC} & \frac{1}{C} & 0 \\ \frac{-1}{L} & 0 & 0 \\ 1 & 0 & 0 \end{pmatrix}, \quad \mathbf{B}_1 = \begin{pmatrix} 0 \\ \frac{V_g}{L} \\ V_{ref} \end{pmatrix}, \tag{A.9}$$

$$\mathbf{A}_3 = \begin{pmatrix} \frac{-1}{RC} & 0 & 0 \\ 0 & 0 & 0 \\ 1 & 0 & 0 \end{pmatrix}, \quad \mathbf{B}_2 = \mathbf{B}_3 = \begin{pmatrix} 0 \\ 0 \\ V_{ref} \end{pmatrix} \tag{A.10}$$

Due to the additional configuration, the discrete-time model of a converter working in DCM, requires revisiting the terms of the Jacobian (Eq. 3.18):

$$\mathbf{DP} = \mathbf{J_x} - \mathbf{J_d}\mathbf{J}_{\sigma_\mathbf{d}}^{-1}\mathbf{J}_{\sigma_\mathbf{x}} = \Phi(\mathbf{D_2T})(\mathbf{I} - \mathbf{J}_2)\Phi(\mathbf{T} - \mathbf{D_2T})(\mathbf{I} - \mathbf{J}_1)\Phi(\mathbf{D_1T}) \tag{A.11}$$

where D_1 is the duty cycle of the PWM control signal and D_2T is the period of time in which the inductor current is discharged, being $\mathbf{J}_1$ and $\mathbf{J}_2$:

$$\mathbf{J}_1 = \frac{\Delta\dot{\mathbf{x}}\mathbf{K}}{\mathbf{K}\dot{\mathbf{x}}(D_1T^-) - m_c} \tag{A.12}$$

$$\mathbf{J}_2 = \frac{\Delta\dot{\mathbf{x}}_2\mathbf{K}_2}{\mathbf{K}_2\dot{\mathbf{x}}(D_1T + D_2T^-)} \tag{A.13}$$

where

$$\Delta\dot{\mathbf{x}} = \dot{\mathbf{x}}(D_1T^-) - \dot{\mathbf{x}}(D_1T^+) \tag{A.14}$$

$$\Delta\dot{\mathbf{x}}_2 = \dot{\mathbf{x}}(D_1T + D_2T^-) - \dot{\mathbf{x}}(D_1T + D_2T^+) \tag{A.15}$$

and, finally:

$$\mathbf{K} = [k_v \quad k_i \quad k_I] \tag{A.16}$$

$$\mathbf{K}_2 = [0 \quad 1 \quad 0] \tag{A.17}$$

A.4 Notch/Narrowband Amplifier Chaos Controller in a VMC Buck Converter with a PI Compensator

The generic transfer function encompassing both notch and narrowband amplifiers is:

$$G_{c,notch}(s) = \frac{s^2 + 2\xi_1\omega_n + \omega_n^2}{s^2 + 2\xi_2\omega_n + \omega_n^2} \tag{A.18}$$

Note that depending the value of ξ_1 and ξ_2 the transfer function can attenuate ($\xi_1 < \xi_2$) or amplify ($\xi_1 > \xi_2$) at the center frequency ω_n. The state matrices for buck converter with this controller are given by:

$$\mathbf{A}_1 = \mathbf{A}_2 = \begin{pmatrix} \frac{-1}{RC} & \frac{1}{C} & 0 & 0 & 0 \\ \frac{-1}{L} & 0 & 0 & 0 & 0 \\ 0 & 0 & 0 & \omega_n^2 & 0 \\ 1 & 0 & 1 & -2\xi_2\omega_n & 0 \\ -k_I & 0 & 0 & k_I(\xi_2 - \xi_1)2\omega_n & 0 \end{pmatrix} \tag{A.19}$$

$$\mathbf{B}_1 = \begin{pmatrix} 0 \\ \frac{V_g}{L} \\ 0 \\ 0 \\ k_I V_{ref} \end{pmatrix}, \quad \mathbf{B}_2 = \begin{pmatrix} 0 \\ 0 \\ 0 \\ 0 \\ k_I V_{ref} \end{pmatrix} \tag{A.20}$$

and the feedback gains vector can be expressed as:

$$\mathbf{K} = \begin{bmatrix} -k_p & 0 & 0 & k_p(\xi_2 - \xi_1)2\omega_n & 1 \end{bmatrix} \tag{A.21}$$

A.5 High Impedance Output Filter and Narrowband Amplifier FSI Controller in a VMC Buck Converter with a PI Compensator

This system consists of an output filter with high impedance at the switching frequency, formed by a $L_n - C_n$ parallel tank, and a NBA chaos controller in the feedback loop, both centered at the switching frequency as it is shown in Fig. 4.22.. The state matrices of the system under this controller are:

$$\mathbf{A}_1 = \mathbf{A}_2 = \begin{pmatrix} \frac{-1}{RC} & \frac{1}{C} & 0 & 0 & \frac{1}{RC} & 0 & 0 \\ \frac{-1}{L} & 0 & 0 & 0 & 0 & 0 & 0 \\ 0 & 0 & 0 & \omega_n^2 & 0 & 0 & 0 \\ 1 & 0 & 1 & -2\xi_2\omega_n & 0 & 0 & 0 \\ \frac{1}{RC_n} & 0 & 0 & 0 & \frac{-1}{RC_n} & \frac{-1}{C_n} & 0 \\ 0 & 0 & 0 & 0 & \frac{1}{L_n} & 0 & 0 \\ -k_I & 0 & 0 & k_I(\xi_2 - \xi_1)2\omega_n & 0 & 0 & 0 \end{pmatrix} \tag{A.22}$$

$$\mathbf{B}_1 = \begin{pmatrix} 0 \\ \frac{V_g}{L} \\ 0 \\ 0 \\ 0 \\ 0 \\ k_I V_{ref} \end{pmatrix}, \quad \mathbf{B}_2 = \begin{pmatrix} 0 \\ 0 \\ 0 \\ 0 \\ 0 \\ 0 \\ 0 \\ k_I V_{ref} \end{pmatrix} \tag{A.23}$$

The feedback gains vector can be expressed as:

$$\mathbf{K} = [-k_p \quad 0 \quad 0 \quad k_p(\xi_2 - \xi_1)2\omega_n \quad 0 \quad 0 \quad 1] \tag{A.24}$$

A.6 LC Divider in a VMC Buck Converter Under a PI Compensator

The circuit is based on adding a output LC tuned network, as it is shown in Fig. 4.25. The state matrices of the system under this controller are:

$$\mathbf{A}_1 = \mathbf{A}_2 = \begin{pmatrix} \frac{-1}{RC} & \frac{1}{C} & 0 & \frac{1}{C} & 0 \\ \frac{-1}{L} & 0 & 0 & 0 & 0 \\ 0 & 0 & 0 & \frac{1}{C_n} & 0 \\ \frac{1}{L_n} & 0 & \frac{-1}{L_n} & 0 & 0 \\ 0 & 0 & -k_I & 0 & 0 \end{pmatrix} \tag{A.25}$$

$$\mathbf{B}_1 = \begin{pmatrix} 0 \\ \frac{V_g}{L} \\ 0 \\ 0 \\ k_I V_{ref} \end{pmatrix}, \quad \mathbf{B}_2 = \begin{pmatrix} 0 \\ 0 \\ 0 \\ 0 \\ k_I V_{ref} \end{pmatrix} \tag{A.26}$$

And the feedback gains vector can be expressed as:

$$\mathbf{K} = [0 \quad -k_i \quad k_p \quad 0 \quad 1] \tag{A.27}$$

Appendix B
Equivalence of Dynamics in Buck Switching Converters

This section justifies the equivalence of the dynamic behavior of different buck converters with the same value of the dimensionless parameters τ and Γ but with different physical parameters, R, L, C, T, V_g, hence allowing to reduce the number of parameters in the design-space. The frequency response of the RLC second order filter can be described as:

$$G_p(s) = \frac{\omega_0^2}{s^2 + \omega_{RC}s + \omega_0^2} \tag{B.1}$$

Therefore, the complete dynamics of the power plant can be characterized by the two parameters ω_0 and ω_{RC}. Note that the buck converter, shown in Fig. 3.1, is equivalent to a second order low pass filter with a square wave signal (the diode voltage v_d) at its input which can be described with a particular switching frequency $\omega_s = 2\pi f_s$. Let us define $\tau = f_s/\omega_{RC} = RCf_s$ and $\Gamma = f_s^2/\omega_0^2 = LCf_s^2$.

The definition of these new parameters reduces the dimensions of the design parameter space from five to three, namely, Γ, τ and D, instead of the physical

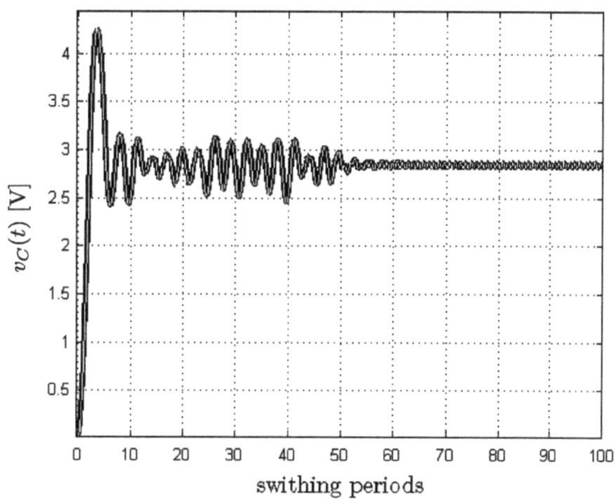

Fig. B.1 Transient simulation of the buck switched model using different values of physical parameters but the same values of dimensionless parameters

E. Rodríguez Vilamitjana et al., *Chaos in Switching Converters for Power Management*,
DOI: 10.1007/978-1-4614-2128-3, © Springer Science+Business Media New York 2013

parameters L, C, R, f_s and D. Note that Γ relates the switching frequency and the natural frequency of the system and τ is related with the output load of the converter. Two converters will be dynamically equivalent, despite they have different physical parameter values, whenever they have the same values of parameters Γ, τ and D as it is exemplified in Fig. B.1 which is obtained for the set of parameter values used in numerical simulations, and those used in the experimental prototype in this work.

Appendix C
Relationship Between the Time Derivative and the Steady State Ripple of the Control Signal

In this appendix it will be demonstrated that the ripple amplitude of the control signal is an indirect estimate of its derivative at the switching instant. In a voltage-mode controlled buck converter, the derivative of the feedback state variable (capacitor voltage), according to Fig. C.1, is

$$\dot{v}_C(DT^-) = \frac{i_C(DT^-)}{C} = \frac{I_{C,max}}{C} \tag{C.1}$$

and its ripple can be expressed as

$$\hat{V}_C = V_{C,max} - V_{C,min} = 1C \int_{t_1}^{t_2} i_C(t)dt = \frac{(t_2 - t_1)I_{C,max}}{2C} \tag{C.2}$$

Note that in the previous equation, it is assumed that the waveform of the capacitor current i_C is triangular. The time duration $t_2 - t_1$ in Eq. (C.2) can be obtained by analyzing the following equations

$$\frac{V_g - \bar{V}_C}{L} t_1 = \frac{\bar{V}_C}{L} t_2 \Rightarrow t_2 = \frac{1 - D}{D} t_1 \tag{C.3}$$

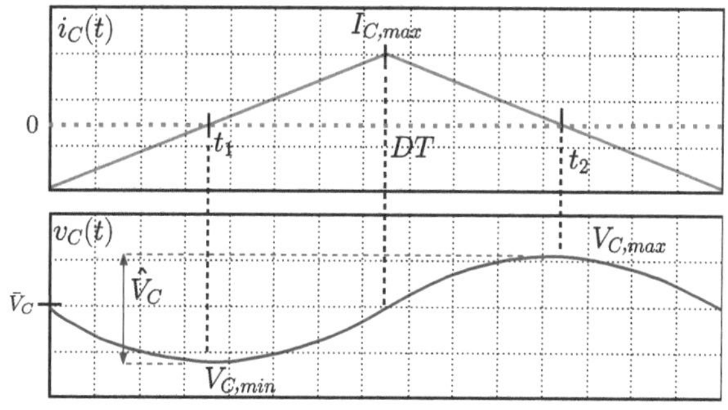

Fig. C.1 Ideal representation of the time domain inductor current i_L (*top*) and output capacitor voltage v_C (*bottom*) in a buck converter

E. Rodríguez Vilamitjana et al., *Chaos in Switching Converters for Power Management*, DOI: 10.1007/978-1-4614-2128-3, © Springer Science+Business Media New York 2013

Fig. C.2 Comparison between the ripple amplitude $\hat{V}_C$ in a voltage-mode controlled buck converter by sweeping Γ and τ.**a** From Eq. (D.5) (*white*) and normalized derivative at the switching instant of feedback state variable $\dot{v}_C(DT^-)(4f_s)^{-1}$ (*black*),**b** error between both results

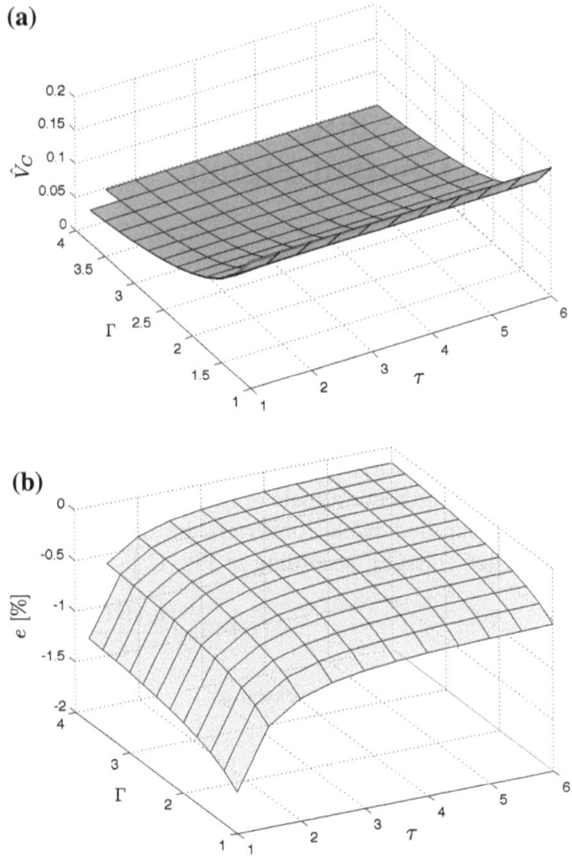

$$\frac{V_g - \bar{V}_C}{2L} DT = \frac{V_g - \bar{V}_C}{L} t_1 \Rightarrow t_1 = \frac{DT}{2} \tag{C.4}$$

$$t_2 - t_1 = T4 = 14f_s \tag{C.5}$$

where $\bar{V}_C$ is the average output voltage. Equation (C.3) assumes steady-state while Eq. (C.4) assumes zero average capacitor current. Therefore, taking into account the expression of the state-variable derivative in Eqs. (C.1), (C.2) and (C.5):

$$\dot{v_C}(DT^-) = 4f_s\hat{V}_C \tag{C.6}$$

Figure C.2 validates the approximation by calculating the derivative and the ripple from the exact waveforms, obtained from an exact Laplace analysis of the output ripple, developed in Appendix D, and varying the parameters τ and Γ. Note that the steady state error increases when τ is relatively low, hence losing the validity of the approach.

Appendix D
Exact Expression of the Output Voltage Ripple for the Buck Converter Derived from Laplace Analysis

A closed-form output voltage ripple expression for the buck converter can be derived using the Laplace transform. The output voltage signal is the result of applying the periodic square wave diode voltage v_d with duty cycle D and amplitude V_g to the RLC second order filter $G_p(s)$ (B.1). Therefore, the Laplace transform $V_C(s)$ of the output voltage $v_C(t)$ is given by the product of the transfer function $G_p(s)$ and the Laplace transform $V_d(s)$ of the diode voltage $v_d(t)$:

$$V_C(s) = V_d(s)G_p(s) = \frac{(1-e^{-sDT})V_g}{s(1-e^{-sT})}\frac{\omega_0^2}{s^2+2\xi\omega_0 s+\omega_0^2} \tag{D.1}$$

where $\xi = \frac{\omega_{RC}}{2\omega_0}$. The partial fraction decomposition of $V_C(s)$ is:

$$V_C(s) = \frac{k_1}{s} + \frac{k_2}{s-a_1} + \frac{k_3}{s-a_2} + \frac{P_0(s)}{1-e^{-sT}} \tag{D.2}$$

where $P_0(s)$ corresponds to the steady-state response and therefore includes the exact output voltage ripple for the buck converter:

$$P_0(s) = (1-e^{-sT})\left(V_C(s) - \frac{k_1}{s} + \frac{k_2}{s-a_1} + \frac{k_3}{s-a_2}\right) \tag{D.3}$$

with $a_1 = -\omega_{RC}/2 - j\omega_d$, $a_2 = -\omega_{RC}/2 + j\omega_d$, $k_1 = D$ and k_2 and k_3 are given by

$$k_2 = \frac{(1-e^{-a_1 DT})\omega_0^2}{2a_1(1-e^{-a_1 T})j\omega_d}, k_3 = k_2^* \tag{D.4}$$

being $\omega_d = \omega_0\sqrt{1-\xi^2}$. Finally the exact time-domain expression of the steady-state output voltage ripple $v_C^*(t)$ for the buck converter operating in CCM can be written in the following form

E. Rodríguez Vilamitjana et al., *Chaos in Switching Converters for Power Management*, DOI: 10.1007/978-1-4614-2128-3,

Fig. D.1 Theoretical waveforms of the output voltage $v_C^*(t)$ of a buck converter obtained from Eq. (D.6) (with an additional DC voltage V_{ref}) and from simulating the switched model in open-loop. $V_g = 6\,\text{V}$, $V_{ref} = 3\,\text{V}$, $R = 2.5\,\Omega$, $L = 66\,\text{nH}$, $C = 20\,\text{nF}$, $f_s = 50\,\text{MHz}$

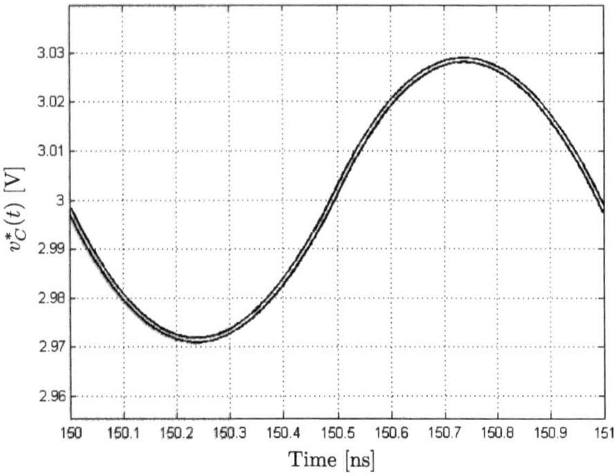

$$
\begin{aligned}
v_C^*(t) = P_0(t) = u(t)V_g\Bigg[1 - e^{-\omega_{RC}t/2}\left(\cos(\omega_d t) + \frac{\xi}{1-\xi^2}\sin(\omega_d t)\right) \\
- D - 2e^{-\omega_{RC}t/2}(\alpha\cos(\omega_d t) - \gamma\sin(\omega_d t))\Bigg] \\
- u(t-DT)V_g\Bigg[1- \\
- e^{-\omega_{RC}(t-DT)/2}\left(\cos(\omega_d(t-DT)) + \frac{\xi}{1-\xi^2}\sin(\omega_d(t-DT))\right)\Bigg]
\end{aligned}
\tag{D.5}
$$

where $u(t)$ is unit step function, $\alpha = \Re(k_2)$ and $\gamma = \Im(k_2)$. Alternatively, the expression can be also developed as:

$$
\begin{aligned}
v_C^*(t) = u(t)V_g\Bigg[1 - e^{-\omega_{RC}t/2}\left(\cos(\omega_d t) + \frac{\xi}{1-\xi^2}\sin(\omega_d t)\right) \\
- D + \mathcal{K}e^{-\omega_{RC}t/2}\Bigg[\left(\frac{\omega_{RC}}{2}\sin(\omega_d t) + \omega_d\cos(\omega_d t)\right) \\
- e^{\omega_{RC}T/2}\left(\frac{\omega_{RC}}{2}\sin(\omega_d(t+T)) + \omega_d\cos(\omega_d(t+T))\right) \\
- e^{\omega_{RC}(DT+T)/2}\left(\frac{\omega_{RC}}{2}\sin(\omega_d(t+(1-D)T) + \omega_d\cos(\omega_d(t+(1-D)T)\right)\Bigg]\Bigg] \\
- u(t-DT)V_g\Bigg[1- \\
- e^{-\omega_{RC}(t-DT)/2}\left(\cos(\omega_d(t-DT)) + \frac{\xi}{1-\xi^2}\sin(\omega_d(t-DT))\right)\Bigg]
\end{aligned}
\tag{D.6}
$$

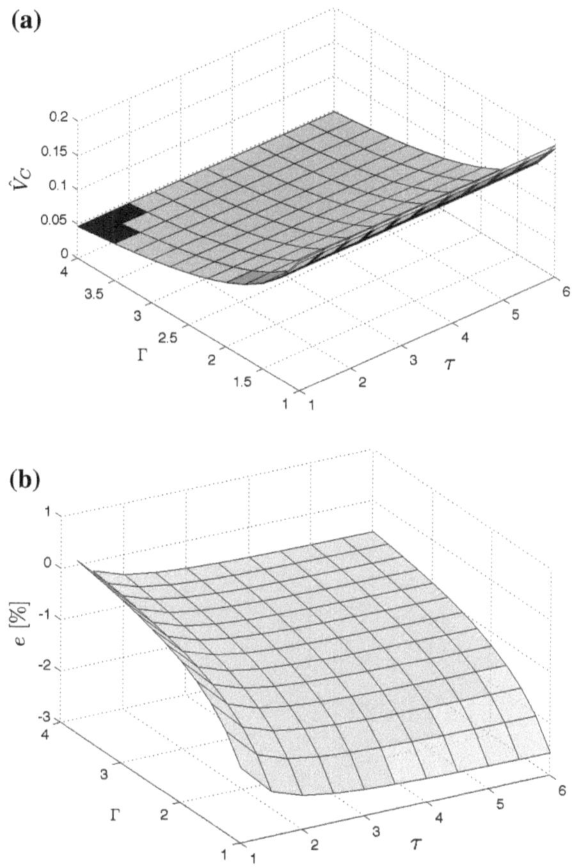

Fig. D.2 **a** Voltage ripple amplitude $\hat{V}_C$ from the exact Laplace expression ((D.5) *black*) and from the approximated expression in Eq. ((3.11) *white*) and **b** Error between both results

where $\mathcal{K} = (\omega_d - 2\omega_d e^{\omega_{RC}T/2}\cos(\omega_d T) + e^{\omega_{RC}T})^{-1}$.

This closed-form expression has been validated by comparing it with the output voltage waveform obtained from numerical simulation of the switched model for an ideal buck converter, as shown in Fig. D.1 The waveforms are coincident, thus validating the expression given in Eq. (D.6) as an exact description of the converter output ripple for all conditions, as opposite to conventional expressions valid only under low ripple conditions. From such exact ripple expression it is possible to compare the accuracy of the ripple approximation proposed in Eq. (3.11). The result of this comparison is shown in Fig. D.2, in which it is possible to observe that the approximation loses validity for low values of τ and Γ.

References

1. Ahlborn A, Parlitz U (2006) Chaos control using notch filter feedback. Phys Rev Lett 96:3
2. Ahn C, Ahn C, Allen M (1996) A comparison of two micromachined inductors (bar- and meander-type) for fully integrated boost DC-DC power converters. IEEE Trans Power Electron 11:239–245
3. Alarcon E, Villar G, Ferrandez S, Guinjoan F, Poveda A (2004) Ripple-reduction tuned filtering switching power converter topology. In: IEEE power electronics specialists conference (PESC'04)
4. Allard B, Trochut S, Lin-Shi X, Retif JM (2004) Control design for integrated switch-mode power supplies: a new challenge? In: IEEE power electronics specialists conference (PESC'04), vol 6, pp 4492–4497
5. Angulo F, Burgos J, Olivar G (2007) Chaos stabilization with TDAS and FPIC in a buck converter controlled by lateral PWM and ZAD. In: Mediterranean conference on control automation MED '07, 2007, pp 1–6
6. Aparicio R, Hajimiri A (2002) Capacity limits and matching properties of integrated capacitors. IEEE J Solid-State Circ 37(3):384–393
7. Athalye A, Grantham W (1995) Notch filter feedback control of a chaotic system. In: Proceedings of the American control conference, pp 837–841
8. Banerjee S, Chakrabarty K (1998) Nonlinear modeling and bifurcations in the boost converter. IEEE Trans Power Electron 13(2):252–260
9. Banerjee S, Karthik MS, Yuan G, Yorke JA (2000) Bifurcations in one-dimensional piecewise smooth maps: theory and applications in switching circuits. IEEE Trans Circ Syst I Fundam Theory Appl 47(3):389–394
10. Banerjee S, Ranjan P, Grebogi C (2000) Bifurcations in two-dimensional piecewise smooth maps: theory and applications in switching circuits. IEEE Trans Circ Syst I Fundam Theory Appl 47(5):633–643
11. Banerjee S, Kastha D, SenGupta S (2002) Minimising EMI problems with chaos. In: International conference on electromagnetic interference and compatibility, pp 162–167
12. Banerjee S, Parui S, Gupta A (2004) Dynamical effects of missed switching in current-mode controlled dc-dc converters. IEEE Trans Circ Syst II Express Briefs 51(12):649–654
13. Batchvarov JS, Valchev VC, Yudov DD, Duarte JL (2002) Investigation of chaos in interleaved power converters. In: International IEEE symposium intelligent systems, pp 79–83
14. Batlle C, Fossas E, Olivar G (1997) Time-delay stabilization of the buck converter. In: International conference control of oscillations and chaos, pp 590–593
15. Bergveld H, Karadi R, Nowak K (2008) An inductive down converter system-in-package for integrated power management in battery-powered applications. In: IEEE power electronics specialists conference (PESC'08), (2008), pp 3335–3341

E. Rodríguez Vilamitjana et al., *Chaos in Switching Converters for Power Management*,
DOI: 10.1007/978-1-4614-2128-3,

16. Blakley J, Illing L, Gauthier DJ (2004) Controlling fast chaos in delay dynamical systems. Phys. Rev. Lett 92(19):193901
17. Brown A, Middlebrook R (1981) Sampled-data modeling of switching regulators. In: IEEE power electronics specialists conference (PESC'81), (1981)
18. Bryant B, Kazimierczuk M (2005) Modeling the closed-current loop of PWM boost DC-DC converters operating in CCM with peak current-mode control. IEEE Trans Circ Syst I Regul Papers 52(11):2404–2412
19. Calvente J, Guinjoan F, Martinez L, Poveda A (1996) Subharmonics, bifurcations and chaos in a sliding-mode controlled boost switching regulator. In: International symposium on circuits and systems (ISCAS'96), (1996), pp 573–576
20. Chakrabarty K, Poddar G, Banerjee S (1996) Bifurcation behavior of the buck converter. IEEE Trans Power Electron 11(3):439–447
21. Chen G, Yu X (1999) On time-delayed feedback control of chaotic systems. IEEE Trans Circ Syst I Fundam Theory Appl 46(6):767–772
22. Chen G, Chen G, Dong X (1993) Control of chaos-a survey. In: IEEE conference on decision and control, pp 469–474
23. Cheng K, Liu M, Ho YL (2003) Experimental confirmation of frequency correlation for bifurcation in current-mode controlled buck-boost converters. IEEE Power Electron Lett 1(4):101–103
24. Corradini L, Mattavelli P (2008) Modeling of multisampled pulse width modulators for digitally controlled DC-DC converters. IEEE Trans Power Electron 23(4):1839–1847
25. Corradini L, Mattavelli P, Tedeschi E, Trevisan D (2008) High-bandwidth multisampled digitally controlled DC-DC converters using ripple compensationt. IEEE Trans Ind Electron 55(4):1501–1508
26. Corradini L, Bjeletic A, Zane R, Maksimovič D (2009) Fully digital hysteretic modulator for DC-DC switching converters. In: IEEE energy conversion congress and exposition (ECCE'09), (2009) pp 3312–3319
27. Dai D, Li S, Ma X, Tse CK (2007) Slow-scale instability of single-stage power-factor-correction power supplies. IEEE Trans Circ Syst I Regul Papers 54(8):1724–1735
28. Dancy A, Chandrakasan A (1997) Ultra low power control circuits for PWM converters. In: IEEE power electronics specialists conference (PESC'97), (1997), pp 21–27
29. Deane JHB (1992) Chaos in a current-mode controlled boost DC-DC converter. IEEE Trans Circ Syst I Fundam Theory Appl 39(8):680–683
30. Deane JHB, Hamill DC (1989) Instability, subharmonics and chaos in power electronic systems. In: IEEE power electronics specialists conference (PESC'89), (1989), pp 34–42
31. Deane JHB, Hamill DC (1990) Analysis, simulation and experimental study of chaos in the buck converter. In: IEEE power electronics specialists conference (PESC'90), pp 491–498
32. Deane JHB, Hamill DC (1991) Chaotic behaviour in current-mode controlled DC-DC convertor. IET Electron Lett 27:1172–1173
33. Deane JHB, Hamill DC (1996) Improvement of power supply EMC by chaos. IET Electron Lett 32:1045
34. Deane JHB, Ashwin P, Hamill DC, Jefferies DJ (1999) Calculation of the periodic spectral components in a chaotic DC-DC converter. Trans Circ Syst I: Fundam Theory Appl 46(11):1313–1319
35. Deisch CW (1978) Simple switching control method changes power converter into a current source. In: IEEE power electronics specialists conference (PESC'78), pp 300–306
36. di Bernardo M, Vasca F (2000) Discrete-time maps for the analysis of bifurcations and chaos in DC/DC converters. IEEE Trans Circ Syst I Fundam Theory Appl 47(2):130–143
37. di Bernardo M, Garefalo F, Glielmo L, Vasca F (1997) Analysis of chaotic buck, boost and buck-boost converters through switching maps. In: IEEE power electronics specialists conference, (PESC '97), pp 754–760
38. di Bernardo M, Garefalo F, Glielmo L, Vasca F (1998) Switchings, bifurcations, and chaos in DC/DC converters. IEEE Trans Circ Syst I Fundam Theory Appl 45(2):133–141

39. El Aroudi A, Robert B (2006) Discrete time model of a multi-cell DC-DC converter: non linear approach. Math Comput Simul J 71:310–319
40. El Aroudi A, Benadero L, Toribio E, Olivar G (1999) Hopf bifurcation and chaos from torus breakdown in a PWM voltage-controlled DC-DC boost converter. IEEE Trans Circ Syst I Fundam Theory Appl 46(11):1374–1382
41. El Aroudi A, Debaat M, Giral L, Olivar G, Benadero L, Toribio E (2005) Bifurcations in DC-DC swithching converters: reivew of methods and applications. Int J Bifurcation Chaos (IJBC'05) 15(5):1549–1578
42. El Aroudi A, Orabi M, Martinez-Salamero L (2008) A representative discrete-time model for uncovering slow and fast-scale instabilities in boost powe factor correlation AC-DC pre-regulators. Int J Bifurcation Chaos (IJBC'08) 18:3073–3092
43. El Aroudi A, Alarcon E, Rodriguez E, Robert B (2008) A qualitative comparison of bifurcation in single-cell and two-cell buck converter. In: 5th International multi-conference on systems, signals and devices (SSD'08), (2008)
44. Erickson RW, Maksimovič D (2001) Fundamentals of Power Electronics. Kluwer, Boston
45. Escobar G, Martinez P, Leyva-Ramos J, Mattavelli P (2006) A negative feedback repetitive control scheme for harmonic compensation. IEEE Trans Ind Electron 53(4):1383–1386
46. Fang CC, Abed E (2001) Harmonic balance analysis and control of period doubling bifurcation in buck converters. IEEE Int Symp Circ Syst (ISCAS'01) 2:209–212
47. Fossas E, Olivar G (1996) Study of chaos in the buck converter. IEEE Trans Circ Syst I Fundam Theory Appl 43(1):13–25
48. Giaouris G, Maity S, Banerjee S, Pickert V, Zahawi B (2009) Application of Filippov method for the analysis of subharmonic instability in DC-DC converters. Int J Circ Theory Appl 37:899–919
49. Giral R, El Aroudi A, Martinez-Salamero L, Leyva R, Maixe J (2001) Current control technique for improving EMC in power converters. Electron Lett 37:274–275
50. Hamill DC, Jeffries DJ (1988) Subharmonics and chaos in a controlled switched-mode power converter. IEEE Trans Circ Syst 35(8):1059–1061
51. Hamill DC, Deane JHB, Jefferies DJ (1992) Modeling of chaotic DC-DC converters by iterated nonlinear mappings. IEEE Trans Power Electron 7(1):25–36
52. Hamill D, Deane J, Aston P (1997) Some applications of chaos in power converters. In: IEEE colloquium on update on new power electronic techniques, pp 1–5
53. Huang Y, Iu HHC, Tse CK (2007) Boundaries between fast-and slow-scale bifurcations in parallel-connected buck converters. In: IEEE international symposium on circuits and systems (ISCAS'07), 2007, pp 2419–2422
54. Iu H, Tse CK (2000) Bifurcation behavior in parallel-connected buck converters. IEEE Trans Circ Syst I Fundam Theory Appl 48(2):233–240
55. Iu H, Tse C (2005) Comparative study of bifurcation in single and parallel-connected buck converters under current-mode control: disapperance of period-doubling. Circ Syst Signal process 24:201–219
56. Iyengar S, Iyengar S, Liakopoulos T, Ahn C (1999) A DC-DC boost converter toward fully on-chip integration using new micromachined planar inductors. In: IEEE power electronics specialists conference (PESC'99), 1999, pp 72–76
57. Jung SH, Jung SH, Jung NS, Hwang JT, Cho GH (1999) An integrated CMOS DC-DC converter for battery-operated systems. In: IEEE power electronics specialists conference (PESC'99), pp 43–47
58. Kaouba K, Pelaez-Restrepo J, Feki M, Robert B, El Aroudi A (2012) Improved static and dynamic performances of a two-cell DC-DC buck converter using a digital dynamic time-delayed control. Int J Circuit theory Appl 40:395–407
59. Krein P, Bass R (1990) Types of instability encountered in simple power electronic circuits: unboundedness, chattering, and chaos. In: Applied power electronics conference and exposition (APEC'90), pp 191–194

60. Kursun V, Narendra S, De V, Friedman E (2004) Low-voltage-swing monolithic DC-DC conversion. IEEE Trans Circ Syst II Express Briefs 51(5):241–248
61. Lau BYB (1987) Small-signal frequency response theory for ideal DC-to-DC converter systems, Ph.D. dissertation, California Institute of Technology, 1987. http://resolver.caltech.edu/CaltechETD:etd-03012008-134552
62. Lehman B, Bass R (1996) Switching frequency dependent averaged models for PWM DC-DC converters. IEEE Trans Power Electron 11(1):89–98
63. Li M, Dai D, Ma X, Iu H (2008) Fast-scale period-doubling bifurcation in voltage-mode controlled full-bridge inverter. In: IEEE international symposium on circuits and systems (ISCAS'08), pp 2829–2832
64. Ma Y, Kawakami H, Tse C (2004) Analysis of bifurcation in switched dynamical systems with periodically moving borders: application to power converters. In: IEEE international symposium on circuits and systems (ISCAS'04) vol 4, pp 101–104
65. Magauer A, Banerjee S (2000) Bifurcations and chaos in the tolerance band PWM technique. IEEE Trans Circ Syst I Fundam Theory Appl 47(2):254–259
66. Maity S, Tripathy D, Bhattacharya TK, Banerjee S (2007) Bifurcation analysis of PWM-1 voltage-mode-controlled buck converter using the exact discrete model. IEEE Trans Circ Syst I Regul Papers 54(5):1120–1130
67. Maksimovič D, Zane R (2007) Small-signal discrete-time modeling of digitally controlled PWM converters. IEEE Trans Power Electron 22(6):2552–2556
68. Marco L, Poveda A, Alarcon E, Maksimovic D (2006) Bandwidth limits in PWM switching amplifiers. In: IEEE international symposium on circuits and systems (ISCAS'06), (2006)
69. Mazumder SK, Nayfeh AH, Boroyevich D (2001) Theoretical and experimental investigation of the fast- and slow-scale instabilities of a DC-DC converter. IEEE Trans Power Electron 16(2):201–216
70. Meynard T, Foch H, Thomas P, Courault J, Jakob R, Nahrstaedt M (2002) Multicell converters: basic concepts and industry applications. IEEE Trans Ind Electron 49(5):955–964
71. Middlebrook R, Cuk S (1976) A general unified approach to modelling switching-converter power stages. In: Power electronics specialists conference (PESC'76), pp 18–34 (1976)
72. Mohan S, del Mar Hershenson M, Boyd S, Lee T (1999) Simple accurate expressions for planar spiral inductances. IEEE J Solid-State Circ 34(10):1419–1424
73. Morel C, Bourcerie M, Chapeau-Blondeau F (2004) Extension of chaos anticontrol applied to the improvement of switch-mode power supply electromagnetic compatibility. In: IEEE international symposium on industrial electronics, pp 447–452 (2004)
74. Mukherjee R, Nandi S, Banerjee S (2005) Reduction in spectral peaks of DC-DC converters using chaos-modulated clock. In: IEEE international symposium on circuits and systems (ISCAS'05), pp 3367–3370 (2005)
75. Murdock J (1987) Current mode control arrangement with load dependent ramp signal added to sensed current waveform, Patent US 4,672,518, Technical report, US4672518 (1987)
76. Oppenheim A, Schafer RW, Buck J (1999) Discrete-time signal processing. Prentice Hall, Upper Saddle River
77. Packard DJ (1976) Discrete modeling and analysis of swithcing regulators, Ph.D. dissertation, California Institute of technology
78. Papafotiou G, Margaris N (2002) Instabilities of the buck DC-DC converter under PI control. In: Intelligent systems and control conference (ISC'02), (2002)
79. Parui S, Banerjee S (2003) Bifurcations due to transition from continuous conduction mode to discontinuous conduction mode in the boost converter. IEEE Trans Circ Syst I Fundam Theory Appl 50:1464–1469
80. Pavljasevic S, Maksimovič D (1997) Subharmonic oscillations in converters with current-mode programming under large parameter variations. In: IEEE power electronics specialists conference (PESC'97), pp 1323–1329

81. Peng H, Maksimovic D, Prodic A, Alarcon E (2004) Modeling of quantization effects in digitally controlled DC-DC converters. In: IEEE power electronics specialists conference (PESC 04), pp 4312–4318
82. Peterchev AV, Sanders SR (2003) Quantization resolution and limit cycling in digitally controlled PWM converters. IEEE Trans Power Electron 18:301–308
83. Pettini M, Lima R (1990) Controlling chaos by parametric perturbation. Phys Rev A 66:2545–2548
84. Poddar G, Chakrabarty K, Banerjee S (1995) Control of chaos in the boost converter. Electron Lett 31(11):841–842
85. Prodic A, Maksimovič D, Erickson R (2001) Design and implementation of a digital PWM controller for a high-frequency switching DC-DC power converter. In: IEEE industrial electronics society (IECON'01), pp 893–898
86. Pyragas K (1992) Continuous control of chaos by self-controlling feedback. Phys Lett A 170:421–428
87. Pyragas K (1995) Control of chaos via extended delay feedback. Phys Lett A 206(5):323–330
88. Raab F, Rupp D (1994) High-efficiency single-sideband HF/VHF transmitter based upon envelope elimination and restoration. In: Conference on HF Radio Systems and Techniques
89. Redl R, Novak I (1981) Instabilities in current-mode controlled switching voltage regulators. In: IEEE power electronics specialists conference (PESC'81) pp 17–28
90. Redl R, Sun J (2009) Ripple-based control of switching regulators: an overview. IEEE Trans Power Electron 24(12):2669–2680
91. Ridley RB (1989) A new continuous-time model for current-mode control. IEEE Trans Power Electron 6:382
92. Rodriguez M, Miaja P, Rodriguez A, Sebastian J (2009) Multilevel converter for envelope tracking in RF power amplifiers. In: IEEE energy conversion congress and exposition (ECCE'09), pp 503–510
93. Schrom G, Hazucha P, Hahn J, Gardner D, Bloechel B, Dermer G, Narendra S, Karnik T, De V (2004) A 480-Mhz, multi-phase interleaved buck DC-DC converter with hysteretic control. In: IEEE power electronics specialists conference (PESC'04), vol 6, pp 4702–4707
94. Stanley C, Ronald L (1995) Multi-sectioned power converter having current-sharing controller, Patent EP0693818A2 Applicant: LORAL SPACE SYSTEMS INC (US), Technical Report
95. Stratakos A, Stratakos A, Sanders S, Brodersen R (1994) A low-voltage4 CMOS DC-DC converter for a portable battery-operated system. In: IEEE power electronics specialists conference (PESC'94), pp 619–626
96. Sun C, Lehman B, Sun J (2000) Ripple effects on small signal models in average current mode control. IEEE Appl Power Electron Conf Expo (APEC'00) 2:818–823
97. Sun J, Heck B, Lehman B (2000) Continuous approximation and the stability of averaging. In: Workshop on computers in power electronics (COMPEL'00), pp 139–144
98. Syed A, Ahmed E, Maksimovič D, Alarcon E (2004) Digital pulse width modulator architectures. In: IEEE power electronics specialists conference (PESC'04), pp 4689–4695
99. Tan F, Tan F, Middlebrook R (1995) A unified model for current-programmed converters. IEEE Trans Power Electron 10(4):397–408
100. Toribio E, El Aroudi A, Olivar G, Benadero L (2000) Numerical and experimental study of the region of period-one operation of a PWM boost converter. IEEE Trans Power Electron 15(6):1163–1171
101. Toribio E, Gaston D (2001) Multiple adjustable delay line for electronic systems, Patent ES2159245A1 Applicant: Universitat Politecnica de Catalunya, Technical report

102. Tse CK, Zhou Y, Lau FCM, Qiu SS (2003) Intermittent chaos in switching power supplies due to unintended coupling of spurious signals. In: IEEE power electronics specialist conference (PESC'03), 2003, pp 642–647
103. Tse C (1994) Chaos from a buck switching regulator operating in discontinuous mode. Int J Circ Theory Appl 22(6):263–278
104. Tse C (1994) Flip bifurcation and chaos in three-state boost switching regulators. IEEE Trans Circ Syst I Fundam Theory Appl 41(1):16–23
105. Tse C, Lai Y (2000) Control of bifurcation in current-programmed DC/DC converters: a reexamination of slope compensation. In: IEEE international symposium on circuits and systems (ISCAS'00), 2000, pp 671–674
106. Tse C, Tam C (1994) A quasi-linear controller for DC/DC converter using a TMS320 digital signal processor. In: IEEE power electronics specialists conference (PESC'94), pp 1040–1045
107. Tse C, Dranga O, Iu H (2003) Bifurcation analysis of a power-factor-correction boost converter: uncovering fast-scale instability. In: International symposium on circuits and systems, (ISCAS'03), pp 312–315
108. Tymerski R (1993) Sampled-data modelling of switched circuits, revisited. In: IEEE power electronics specialists conference (PESC '93), pp 39–401
109. Tymerski R, Li D (1993) Extended ripple analysis of PWM DC-to-DC converters. IEEE Trans Power Electron 8(4):588–595
110. Van de Sype DM, De Gusseme K, Van den Bossche AR, Melkebeek JA (2004) Small-signal z-domain analysis of digitally controlled convertersw. In: IEEE power electronics specialists conference (PESC'04), 2004, pp 4299–4305
111. Vasic M, Garcia O, Oliver J, Alou P, Diaz D, Cobos J (2009) Multilevel power supply for high efficiency RF amplifiers. In: IEEE applied power conference and exposition (APEC'09), 2009, pp 1233–2334
112. Verghese GC, Elbuluk M, Kassakian JG (1986) A general approach to sampled-data modeling for power electronics circuits. IEEE Trans Power Electron 1:76–89
113. Verghese G, Bruzos C, Mahabir K (1989) Averaged and sampled-data models for current mode control: a re-examination. IEEE power Electron Specialists Conf (PESC'89) 1:484–491
114. Villar G, Alarcon E (2008) Monolithic integration of a 3-level DCM-operated low-floating-capacitor buck converter for DC-DC step-down donversion in standard CMOS. In: IEEE power electronics specialists conference (PESC'08), pp 4229–4235
115. Villar G, Alarcon E, Martinez H, Biel D, Vidal E, Poveda A (2002) Averaging circuit for switching power converter control: a CMOS current-mode integrated implementation. In: IEEE international symposium on circuits and systems (ISCAS'02), pp 269–272
116. Villar G, Alarcon E, Guinjoan F, Poveda A (2003) A design space exploration for integrated switching power converters. In: IEEE international symposium on circuits and systems (ISCAS'03), pp 304–307
117. Vorperian V (1990) Simplified analysis of PWM converters using model of PWM switch. Continuous conduction mode. IEEE Trans Aerosp Electron Syst 26:490–496
118. Wei-Guo L, Luo-Wei Z, Quan-Ming L (2010) Non-invasive chaos control of DC-DC converter and its optimization. International Journal of Circuit Theory and Applications
119. Wester G (1990) Describing-function analysis of a ripple regulator with slew-rate limits and time delays. In: IEEE power electronics specialists conference (PESC'90), pp 341–346
120. Wester G, Middlebrook R (1972) Low-frequency characterization of a switched DC-DC converters. In: IEEE power electronics specialists conference (PESC'72), pp 9–20
121. Wilkinson R, Meynard T, du Toit Mouton H (2006) Natural balance of multicell converters: the two-cell case. IEEE Trans Power Electron 21:1649–1657
122. Wood JR (1989) Chaos: a real phenomenon in power electronics. In: Applied power electronics conference and exposition (APEC'89), pp 115–124

123. Woywode O, Weber J, Guldner H, Baranovski A, Schwarz W (2003) Bifurcation and statistical analysis of DC-DC converters. IEEE Trans Circ Syst I Fundam Theory Appl 50:1072–1080
124. Wu X, Tse C, Dranga O, Lu J (2005) Fast-scale instability of single-stage power-factor-correction power supplies. In: IEEE international symposium on circuits and systems (ISCAS'05) pp 2477–2480
125. Yousefzadeh V, Alarcon E, Maksimovič D (2005) Three-level buck converter for envelope tracking in RF power amplifiers. In: IEEE Appl Power Electron Conf Expo (APEC'05) 3:1588–1594
126. Yousefzadeh V, Alarcon E, Maksimovič D (2006) Band separation and efficiency optimization in linear-assisted switching power amplifiers. In: IEEE power electronics specialists conference (PESC '06), pp 1–7
127. Yousefzadeh V, Wang N, Popovic Z, Maksimovič D (2006) A digitally controlled DC/DC converter for an RF power amplifier. IEEE Trans Power Electron 21(1):164–172
128. Zhou Y, Iu H, Tse C, Chen JN (2005) Controlling chaos in DC/DC converters using optimal resonant parametric perturbation. In: IEEE international symposium on circuits and systems (ISCAS'05), vol 3, pp 2481–2484

Index

E. Rodríguez Vilamitjana et al., *Chaos in Switching Converters for Power Management*,
DOI: 10.1007/978-1-4614-2128-3,

MIX
Papier aus verantwortungsvollen Quellen
Paper from responsible sources
FSC® C105338

If you have any concerns about our products,
you can contact us on
ProductSafety@springernature.com

In case Publisher is established outside the EU,
the EU authorized representative is:
Springer Nature Customer Service Center GmbH
Europaplatz 3, 69115 Heidelberg, Germany

Printed by Libri Plureos GmbH
in Hamburg, Germany